Michael Steinfeldt

Arnim von Gleich

Ulrich Petschow

Rüdiger Haum

Nanotechnologies, Hazards and Resource Efficiency

Translation sponsored by

Michael Steinfeldt
Arnim von Gleich
Ulrich Petschow
Rüdiger Haum

Nanotechnologies, Hazards and Resource Efficiency

A Three-Tiered Approach to Assessing the Implications of Nanotechnology and Influencing its Development

with 53 Figures

Michael Steinfeldt
University of Bremen
Faculty Production Engineering
FG 10 Technological Design
and Development
Badgasteiner Str. 1
28359 Bremen
Germany

Prof.Dr. Arnim von Gleich
University of Bremen
Faculty Production Engineering
FG 10 Technological Design
and Development
Badgasteiner Str. 1
28359 Bremen
Germany

Ulrich Petschow
Institute for Ecological
Economy Research
Potsdamer Str. 105
10785 Berlin
Germany

Rüdiger Haum
University of Sussex
SPRU-Science and Technology
Policy Research
Falmer
Brighton
UK BN1 9QN

Library of Congress Control Number: 2007931200

ISBN   978-3-540-73882-4 Springer Berlin Heidelberg New York

This work is subject to copyright. All rights are reserved, whether the whole or part of the material is concerned, specifically the rights of translation, reprinting, reuse of illustrations, recitation, broadcasting, reproduction on microfilm or in any other way, and storage in data banks. Duplication of this publication or parts thereof is permitted only under the provisions of the German Copyright Law of September 9, 1965, in its current version, and permission for use must always be obtained from Springer-Verlag. Violations are liable to prosecution under the German Copyright Law.

Springer is a part of Springer Science+Business Media
springer.com
© Springer-Verlag Berlin Heidelberg 2007

The use of general descriptive names, registered names, trademarks, etc. in this publication does not imply, even in the absence of a specific statement, that such names are exempt from the relevant protective laws and regulations and therefore free for general use.

Cover design: deblik, Berlin
Production: Almas Schimmel
Typesetting: Camera-ready by authors

Printed on acid-free paper   30/3180/as 5 4 3 2 1 0

Acknowlegdements

We would like to thank the Degussa GmbH for their financial support in making this book. Furthermore, we would like to thank the German Ministry of Education and Research (BMBF) for funding the project "Effects of the Production and Application of Nanotechnology Products on Sustainability" (FK: 16/1504) as the project's final report was updated, extended and developed further into this publication.

The completion of this book was only possible with the help of the translator Kevin Pfeiffer. He has our sincerest gratitude.

The Autors

Preface

Nanotechnology is frequently described as an enabling technology and fundamental innovation,[1] i.e. it is expected to lead to numerous innovative developments in the most diverse fields of technology and areas of application in society and the marketplace. The technology, it is believed, has the potential for far-reaching changes that will eventually affect all areas of life. Such changes will doubtlessly have strong repercussions for society and the environment and bring with them not only the desired and intended effects such as innovations in the form of improvements to products, processes and materials; economic growth; new jobs for skilled workers; relief for the environment; and further steps toward sustainable business, but also unexpected and undesirable side effects and consequences.

With respect to the time spans in which nanotechnology's full potential will presumably unfold, M. C. Roco (2002:5)[2] identified the following stages or generations for industrial prototypes and their commercial exploitation:

- **Past and present**: The "coincidental" use of nanotechnology. Carbon black, for example, has been in use for centuries; more specific, isolated applications (catalysts, composites, etc.) have been in use since the early nineties.
- **First generation**: Passive nanostructures (ca. 2001). Application particularly in the areas of coatings, nanoparticles, bulk materials (nanostructured metals, polymers, and ceramics).
- **Second generation**: Active nanostructures (ca. 2005). Fields of application: particularly in transistors, reinforcing agents, adaptive structures, etc.

---

[1] In the more recent literature, "system innovation" is most often used. For literature on nanotechnology in general; see for example (Bachmann 1998) as well as (National Science and Technology Council 1999) and (NNI 2000).
[2] In place of Roco (2002) other authors and studies, each with its own time scale, could have been named. This is unimportant for our purposes here, as we first and foremost want to call attention to the dynamics of the developments over time.

- **Third generation**: Three-dimensional nanosystems (ca. 2010) with heterogeneous nanocomponents and various assembling techniques.
- **Fourth generation**: Molecular nanosystems (ca. 2020) with heterogeneous molecules, based on biomimetic processes and new design.

It must come as no surprise that the revolutionary potential of nanotechnology has also led to some somewhat extreme judgments. There is, for example, the "radical green vision," in which nanotechnology will help to solve all environmental pollution problems; then there is the radical horror scenario ("grey goo"), according to which all life on earth will be destroyed by nanobots gone wild.[3] Looking beyond these two extremist scenarios, one ultimately finds fully justifiable societal controversies concerning the direction of development and the opportunities and risks associated with the realization of the tremendous technological potential in the nanoscale domain. A great deal of the controversy revolves around the ecological and economic – not to mention health and social – consequences of nanotechnological development and ideas for the future.

Public discourse and the early dissemination of extensive information on technological consequences and the impact of nanotechnology on sustainability are not only advisable, but therefore necessary. In Germany, three projects for an analysis of the innovation and technological potential of nanotechnology on the following topics have accordingly been carried out:

- The Economic Potential of Nanotechnology; Contractor: VDI Zukünftige Technologien Consulting, Düsseldorf; Deutsche Bank Innovationsteam Mikrotechnologie, Berlin
- Nanotechnology and Health; Contractor: Aachener Kompetenzzentrum Medizintechnik (AKM), Institut für Gesundheits- und Sozialforschung Berlin (IGES), among others.
- Effects of the Production and Application of Nanotechnology Products on Sustainability; Contractor: Institut für ökologische Wirtschaftsforschung (IÖW) in cooperation with the Universität Bremen, asmec GmbH, and Nanosolutions GmbH

The project "Effects of the Production and Application of Nanotechnology Products on Sustainability" addresses the current state of materials and technology assessment and attempts to develop this further toward an integrated prospective sustainability assessment. The focus is therefore on the environmental opportunities and risks in this developing technology.

Inasmuch as nanotechnology is a broad, extremely heterogeneous technological field, a generally accepted definition for it still does not exist.

---

[3] See (Joy 2000; etc group 2002)

This study, therefore, is oriented on Paschen et al. (2003) and Basler & Hofmann (2002) and utilizes the following definition of nanotechnology:

Nanotechnology deals with structures in which at least one dimension is smaller than 100 nm. Nanotechnology takes advantage of characteristic effects and phenomena that arise in the transition region between the atomic and mesoscopic levels. Nanotechnology denotes the selective fabrication and/or manipulation of individual nanostructures.

The project takes up two main questions:
1. How can one successfully evaluate the to-be-anticipated effects of a technology still in the making?
2. How can we successfully influence sustainable development in nanotechnology design?

When a technology's applications are still not fully known, and – as in the case of nanotechnology – there are good reasons to assume that new, unknown effects exist, the only significant variable that remains for investigation is the technology itself. It is therefore advisable to turn our attention from the "effects" to the "cause," i.e., to an analysis and characterization of nanotechnology itself. Furthermore, by concentrating on already known concrete applications, sustainability effects can be collected and assessed using a life cycle assessment approach. Therefore the in-depth case studies particularly address applications utilizing nanoparticles and nanostructured surfaces. Furthermore, new technologies do not simply "grow," in a natural manner, but are instead developed by individuals, each acting of their own accord, using the means available to them, and working in recognizable constellations (innovation systems). New technologies are therefore "formed" and in this process the Leitbild (a concept often translated as "exemplary or formative model," "target concept," or "guiding vision") often plays a significant role. With respect to the second question, we therefore consider the possibilities as to whether and to what extent the help of such Leitbilder and other means can influence the development of nanotechnology.

This book is an updated version of our project report and has the following structure: In the chapters that follow, the results of the studies are summarized. In chapter two the methodological basis for a three-stage approach to prospective technology assessment and development of nanotechnology is introduced. In chapters three through five we present the results that were achieved with the help of this method. Chapter six concludes with a summary of the results, preliminary scientific conclusions, and suggestions for further needed research.

Contents

Acknowlegdements V

Preface VII

Contents XI

Authors XV

1   **Summary** 1
    1.1   Characterization of some nanotechnologies 1
    1.2   Evaluation of specific application contexts 5
    1.3   Formative approaches to sustainable nanotechnology ... 10
    1.4   Conclusion and need for action 13

2   **Methodological approaches to the prospective assessment** 17
    2.1   Characterization of technologies 17
        2.1.1   Dealing with the unknown 17
        2.1.2   The "characterization of technologies" as an approach to prospective technology assessment 23
        2.1.3   Technology characterization and types of hazards 25
    2.2   Environmental profile assessments 27
    2.3   Approaches to shaping technological development 32

3   **Technology-specific impacts of nanotechnology** 39
    3.1   Characterization of nanotechnology 39
    3.2   Description and evaluation of production methods 46
        3.2.1   Gas-phase deposition (CVD, PVD) 46
        3.2.2   Flame-Assisted Deposition 48
        3.2.3   Sol-gel process 49
        3.2.4   Precipitation 51
        3.2.5   Molecular molding 51
        3.2.6   Lithography 52
        3.2.7   Self-organization 53

       3.2.8   Qualitative assessment of the various production methods . 54
    3.3   A digression: "nano-visions" and their risk assessment 59

4 Assessment of sustainability effects in the context of specific applications 65
    4.1   Selection of the in-depth case studies 65
    4.2   Case study 1: Eco-efficient nanocoatings 68
        4.2.1   Content, goal, and methods of the case study 68
        4.2.2   Scope of the investigation 69
        4.2.3   Life cycle inventory analysis 84
        4.2.4   Life cycle impact assessment 89
        4.2.5   Case-study Summary 93
    4.3   Case study 2: Nanotechnology Innovation in Styrene Production 93
        4.3.1   Contents, goals and methods of the case study 93
        4.3.2   Scope of the investigation 94
        4.3.3   Scope of investigation and accessibility of data 99
        4.3.4   Life cycle inventory analysis 106
        4.3.5   Life cycle impact assessment 110
        4.3.6   Impediments to the introduction of nanotechnology to the marketplace 111
        4.3.7   Case-study Summary 111
    4.4   Case study 3: Nano-innovations in displays 112
        4.4.1   Contents, goals, and methods 112
        4.4.2   Scope of the investigation 112
        4.4.3   Life cycle inventory analysis 132
        4.4.4   Life cycle impact assessment 136
        4.4.5   Case-study Summary 136
    4.5   Case study 4: Nano-applications in the lighting industry 137
        4.5.1   Contents, goals, and methods 137
        4.5.2   Scope of the investigation 137
        4.5.3   Life cycle inventory analysis 150
        4.5.4   Life cycle impact assessment 155
        4.5.5   Light sources based on quantum dot technology 155
        4.5.6   Case-study Summary 159
    4.6   Case study 5: The risk potential of nano-scale structures 159
        4.6.1   Potential risks of nanotechnology 160
        4.6.2   Nanoscalar titanium dioxide in suntan lotions 171
        4.6.3   Life cycle analysis of nanomaterials 175
        4.6.4   Discussion 178

5 Formative approaches to a sustainable nanotechnology 181
    5.1   Cooperative design approaches 184

  5.1.1 Various Leitbilder for sustainable nanotechnology development 185
  5.1.2 Road maps as information and communication concepts for the design process 195
  5.1.3 Constructive technology assessment and real-time technology assessment as information and communication concepts for design processes 196
  5.1.4 Upstream communication: the integration of citizens/consumers in technological development 202
 5.2 Player-specific approaches to shaping development 203
  5.2.1 Player business: integration of safety, health, and environmental aspects into comprehensive quality management throughout the value chain 203
  5.2.2 Player business: sustainable nanodesign in research and development 204
  5.2.3 Player government: federal regulatory approaches 205

6 Conclusions, the outlook, and need for action 215
 6.1 Results of environmental assessments of selected nanotechnological application contexts 215
 6.2 Technology characterization and context analysis 216
 6.3 Process- and formative model–based design and development 220

Appendix 223

References 251

Index 269

Authors

Michael Steinfeldt is a diploma'd engineer, and is researcher at the University of Bremen, Department of Production Engineering. His principle fields of interest are sustainability assessment, nanotechnology and sustainability, life cycle assessment and environmental management.

Arnim von Gleich is professor for Technological Design and Development at the University of Bremen, Department of Production Engineering. He has published widely in the scientific and policy research literature on genetic engineering, chemical technologies, technical risks, precautionary principle, innovation, green chemistry, bionics, industrial ecology, supply chain management and strategies towards sustainable development in cooperation with enterprises.

Ulrich Petschow is an economist and head of the environmental economics and politics at the Institute for Ecological Economic Research (IOEW). His main focus of research is environmental valuation and methods of technology assessment. Current research themes are Bionics/biomimetics and nanotechnologies.

Rüdiger Haum holds postgraduate degrees Media Studies and Technology Policy. He is currently a doctoral student and part-time researcher at SPRU-Science and Technology Policy Research and also works as a Research Associate at the Environmental Policy Research Centre at the Free University of Berlin.

1 Summary

This project, "Effects of the Production and Application of Nanotechnology Products on Sustainability," was funded by the German Federal Ministry of Education and Research and addresses the current state of materials and technology assessment and attempts to further development toward an integrated prospective sustainability assessment. The focus is on environmental opportunities and risks of the developing nanotechnologies. The project asks two main questions:
1. How can one successfully assess the to-be-anticipated effects of a technology still in the making?
2. How can we successfully influence sustainable development in nanotechnology design?

In order to do justice to the complexity of the problem definition, the following three-step approach to the prospective assessment of technology and the development of nanotechnology was utilized in the project.

- **1st approach – prospective**
  Assessment of nanotechnology and its to-be-expected effects by means of a characterization of the technology.
- **2nd approach – concurrent**
  Evaluation of sustainability effects utilizing life cycle assessment methods (extrapolating) on typical applications in comparison to existing products and processes.
- **3rd approach – formative**
  Leitbilder as "guiding instruments" in technology development, associated processes, actor-specific concepts.

1.1 Characterization of some nanotechnologies

Every form of engineering results assessment has to struggle with the prognosis problem and deal with lack of knowledge (the unknown or unknowable), and uncertainty. The prospective-oriented approach focuses on the assessment of nanotechnology and its anticipatable effects by means of a "characterization of the technology" (described in detail in Gleich 2004). Awareness of and serious consideration of the problem of the unknown in the development of new technology makes it possible with such a characterization to derive and describe possible risks as well as positive effects.

Nanotechnological processes are first and foremost characterized by the dimension in which they take place. In the nanoworld we are at the level of individual molecules and atoms, in a realm measured in millionths of a millimeter. An example of what makes this dimension special is the behavior of nanoparticles, which is generally quite different from that of their more coarse-grained counterparts. For example, the large specific surface area of nanoparticles leads, as a rule, to an increase in chemical reactivity and/or catalytic activity. The relatively small number of atoms in nanoparticles leads, on the other hand, to deviations in optical, electrical, and magnetic properties. Beyond these fundamental characteristics of nanotechnology, other positive effects and potential benefits and/or possible, anticipatable problematic effects can be derived.

Table 1. Characteristics of nanomaterials and thus anticipatable positive ecological benefits or potentials and/or problematic effects[4]

Nano-characteristic		+ Positive environmental impact and benefits / - Problems and hazards	Assessment approaches
Small particle size and particle mobility	+ -	Selective use for resource- and eco-efficient technology Absorbed by the lungs and alveoli Passes through cell membranes, via the olfactory nerve directly into the brain Mobility, persistence and solubility as indicators for bioaccumulation and environmental hazard	Life cycle assessment (LCA), dispersal and exposure models, (eco-) toxicological testing, animal testing, epidemiology
Precision, particle size / layer size, purity	+ -	Selective use for resource- and eco-efficient technology Increased production costs, higher material and energy streams, increased use of resources	LCA, entropy balance, question of "ecological amortization"
Material characteristics	+	Possible replacement for substances dangerous to health and environment	Toxicology, ecotoxicology,

---

[4] Source: Based on (Gleich 2004) and (Steinfeldt 2003)

	−	Health and environmental dangers due to hazardous (rare) elements or substance groups in environmentally open applications	relationship between "natural" and "anthropogenic" material streams
Adhesion, cohesion, agglomeration	+	"Intrinsic safety" due to adhesive, cohesive, and agglomerative tendencies of nanoparticles, thus loosing their nanocharacteristics	Dispersal and exposure models, (environmental) (eco-) toxicology testing, animal testing, epidemiology, atmospheric chemistry, risk analysis
	−	Behavior of nanoparticles or fibers "set free" in the environment Mobilization and inclusion effect of nanoparticles on toxins and heavy metals (piggybacking)	
New chemical effects, modified behavior	+	Utilization of modified behavior for resource and environmentally efficient technology, e.g. use of catalytic effects for more efficient chemical processes or in the environment	LCA, dispersal and exposure models, (eco-) toxicology testing (e.g., allergy / sensitization testing), animal testing, epidemiology, atmospheric chemistry, risk analysis
	−	Changes in: solubility, reactivity, selectivity, catalytic and photocatalytic effects, and temperature dependence of phase transitions mean that surprising technological, chemical, toxicological, and environmentally toxic effects can be expected	
New physical effects, modified optical, electrical, magnetic behavior	+	Selective utilization of effects and modified properties for resource and environmentally efficient technology, e.g. GMR effect, Tyndall effect, quantum effects, tunnel effect	LCA, for technological systems: FMEA, fault-tree analysis
	−	Generally dependent on purified, precisely regulated "technical environments"; there (in the case of non-compliance) surprises can be expected (technical failure)	
Self-organization Self-replication	+	Selective use for resource / eco-efficient and consistent technology	Risk analysis, depth of intervention, LCA, environmental impact analysis, scenario techniques
	−	Danger of uncontrolled developments, self-reproducing nanobiostructures	

This characterization overview makes clear that with the new properties of nanoparticles many hazards could arise, particularly when handled in an open environment. This issue, also of the highest priority in the current discourse on risks, was therefore separately addressed in an in-depth case study.

In an assessment of the hazards of further applications – in part, far in the future – in the area of self-organization (particularly combinations of nanotechnology and biotechnology, and possibly nanotechnology and robotics), the aspect of possible self-replication or reproduction becomes much more significant. However, it must be noted that the potential for self-replication presented by a fusion of nanotechnology and biotechnology or genetic engineering may much more likely than that based on a merger of nanotechnology and robotics. The capability of self-replication such as that possessed by genetically modified organisms may anyhow open up new hazards with respect to health and environmental dangers; whereas this is less of a hazard for pure self-organization. As long as nanotechnology limits itself to dealing with molecules, such a step from the self-organization of molecules to self-replication and copying of organisms (or assemblers) – unless intentionally pursued – is rather unlikely. However, with the merger of nanotechnology and the genetic modification of organisms capable of self-reproduction, such a step could be entirely feasible.

This characterization of nanotechnology at the technological level, which certainly looks ahead to developments far in the future, was followed by a qualitative analysis of currently relevant manufacturing processes for nanoparticles and nanostructured materials in this already particularly well-developed area of nanotechnology, specifically chemical vapor deposition, the Siemens-Martin (open hearth) and sol-gel processes, precipitation, molecular imprinting, lithography, and self-organizing processes – e.g., self-assembled monolayers (SAMs) – with respect to their technological and thereby associated energy and material expenditures as well as their potential risk for the release of nanoparticles.

In contrast to other processes, these all possess a higher potential for the release of nanoparticle emissions in a gaseous form, which could lead to direct emissions in the workplace and the production of loose nanoparticles, for example, in flame-assisted deposition. In other processes, the risks are judged to be rather minimal, assuming that the resulting emissions are adequately handled by an appropriate exhaust air or waste water treatment facility capable of inhibiting the emission of nanoparticles. Furthermore, some of the process technologies presented, such as those that have already been implemented as fundamental technologies in microelectronics and optoelectronics, are notable for an immense technical and energy expenditure yielding a rather small absolute quantitative output.

1.2 Evaluation of specific application contexts

Building on this characterization of specific nanotechnologies and their manufacturing processes to date, the ongoing extrapolative assessment approach tracks the investigation of sustainability effects using specific application examples that are compared to existing products and processes. In this process, the focus was placed on environmental opportunities and risks.

As an assessment approach, the prepared environmental profiles are modeled on the life cycle assessment methodology. The life cycle assessment (LCA) is the most extensively developed and standardized methodology for assessing the environmental aspects and product-specific potential environmental impacts associated with the complete life cycle of a product. It has the advantage that by means of comparative assessments, an (extrapolative) analysis of eco-efficiency potential in comparison to existing applications is possible. With its method of extrapolation, however, this study goes far beyond the current state of the methodology. At the same time, the LCA methodology – as with all methodologies – has its characteristic deficits; there are impact categories for which generally accepted impact models and quantifiable assessments do not exist. This is particularly true in the relevant categories of human and environmental toxicity. And so, a consideration of the impact of fine dust particles (the PM-10 risk, for example, deals with the potential toxicity of particles smaller than 10 μm) in assessments of nanotechnology applications is therefore already doomed to failure because of its reference to material flows expressed in weight. Furthermore, in LCA assessments the risks and the technological power (hazard) effectiveness of applications are not considered. A comprehensive methodology must provide for such analyses.

In the project an attempt was therefore made to compensate for these methodological shortcomings through the establishment of priorities in the selection of the specific application contexts. From the spectrum of all nanotechnology applications, four case studies were chosen in which, on the basis of a preliminary assessment and qualitative evaluation, particularly interesting eco-efficiency potentials could be expected. A further constraint in the selection of the case studies resulted from the requirement that only those examples be considered for which LCA data was (at a minimum) already available for the enlisted "conventional" technologies or products to be used in the comparison. The potential risks and hazards in those nanotechnology applications that were not considered were then analyzed and considered in a separate hazard analysis focusing specifically on nanoparticles.

Table 2. Overview of the case studies investigated

Application context	Goal
Eco-efficient nanocoatings	Presentation of the eco-efficiency potential of nanocoatings in the form of a comparative ecological profile (Nanocoating based on sol-gel technology as compared with waterborne, solventborne, and powder coat industrial coatings)
Nanotechnological process innovation in styrene production	Presentation of the eco-efficiency potential of nanotechnology in catalytic applications in the form of a comparative ecological profile (Nanotube catalyst as compared with iron oxide–based catalysts)
Nanotechnological innovation in the video display field	Assessment of eco-efficiency potential in video display development by means of a qualitative comparison (Organic LED displays and nanotube field emitter displays as compared with CRT, liquid-crystal, and plasma screens)
Nano-applications in the lighting industry	Presentation of eco-efficiency potential of nano-applications in the lighting industry in the form of a comparative ecological profile (White LED and quantum dots as compared with incandescent lamps and compact fluorescents)
Potential risks of nanotechnological applications involving nanoparticles	Discussion of possible risks and harzards using titanium dioxide as an example; less a consideration of environmental impact

The results of the LCA comparisons make clear that nanotechnology applications neither intrinsically nor exclusively can be associated with the potential for a large degree of environmental relief. Nevertheless, for the majority of the application contexts – selected with these aspects in mind – significant eco-efficiency potentials could be ascertained using the chosen methods of comparative analysis of functionality.

The reliability of the ascertained numbers is, of course, dependant on the quality and accessibility of the material and energy data available for the individual applications. For those nanotechnological processes still in development, almost no quantitative evidence is available, although for the usage phase estimates (mostly in energy-savings potential) are possible. Likewise, when comparing established or mature technologies with those still in development, one must recognize that the new technology is at the beginning of its "learning curve," i.e., that it holds the potential for significantly greater increases in efficiency.

The case study "**Eco-efficient Nanocoatings**" makes impressively clear that in the field of surface coatings, with respect to all emissions and environmental considerations studied; there is great potential for a very high degree of eco-efficiency through the utilization of nanotechnology-based coatings. It was also possible to demonstrate the further advantages of a simplified pretreatment process (no chromating). The minimal coating thickness necessary to achieve the same level of functionality makes possible a five-fold increase in resource efficiency. Advantages in the use phase are particularly to be expected in the transport sector as the trend to lighter-weight fabrication continues. In addition to the automotive industry, the potential for greater efficiency will have an even greater effect on the airline and rail industries. A further potential for optimization can be found in the reduction of the proportion of solvents in nanocoating applications.

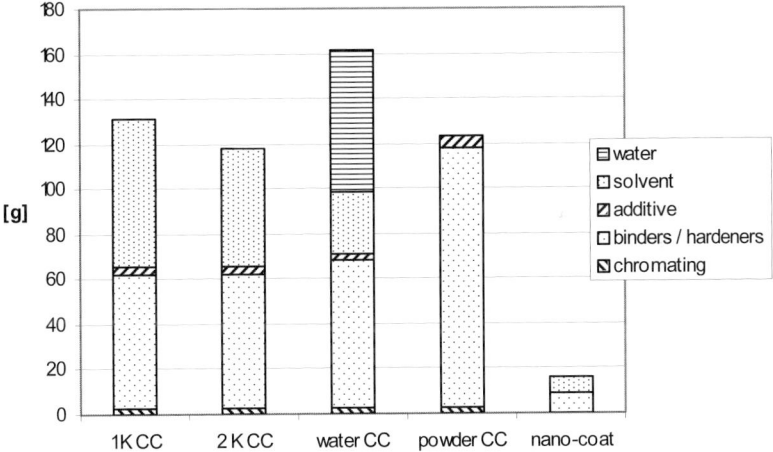

Fig. 1. Coating and chromating quantities (g/m² coated aluminum automobile surface area)[5]

In the course of the case study "**Nanotechnological Process Innovation in the Production of Styrene**," we took as our example for investigation the ecological potential of nanotechnology-based catalytic applications. Specifically, the deployment of a nanostructured catalyst based on nanotubes for the chemical production (synthesis) of styrene. Since no detailed

---

[5] Source: authors (base data: authors, Harsch and Schuckert 1996)

life cycle assessment data for this alternative styrene production process was available, data on the energy use for this process were derived from descriptions of the technology.

The alternative styrene production process, based on a nanotube catalyst, thus yielded potential energy savings of almost 50% at the process stage. Two special effects are responsible for this improvement in efficiency: first, the previously endothermic reaction could be converted to an exothermic one; second, the reaction temperature could be lowered considerably, the reaction medium altered, and the plant power input minimized. With regard to the overall styrene product life cycle, this means an increase in efficiency of about 8–9%. Furthermore, the new catalyst makes possible considerable reductions in heavy metal emissions during the product life cycle. Possible risks from the deployment of nanotubes could negatively offset this; this still needs to be further investigated and taken into consideration in possible facility planning.

The goal of the case study "**Nano-innovations in Display Technology**" was the investigation of possible eco-efficiency potentials in the transition to new nanotechnology-based displays that just now are in the early stages of development. In this case study, OLEDs as well as CNT FEDs were compared to conventional CRT and modern LCD and plasma displays.

This case study likewise suggests that increases in material and energy efficiencies are possible, although due to the various stages of development of the technology, the resulting eco-efficiency estimates come with a certain degree of uncertainty. After overcoming the problem of long-term stability of the organic luminescent materials, the lower production costs of OLEDs as compared to those of the prevailing LCDs could stand in their favor. OLEDs also promise a greater degree of energy efficiency in the use phase (by a factor of two as compared to the LCD). The successful implementation in mass production of these material and energy increases in efficiency will make possible significant eco-efficiency potentials. A minimum twenty-percent savings in life-cycle energy use as compared to the LCD seems possible.

The development of higher eco-efficiency potentials for CNT FEDs will also be possible once energy efficiencies in the manufacturing phase, particularly the highly complex production of nanotubes for field emitters, becomes comparable to current production processes. Risks from these new technologies are unlikely.

Our goal in the case study "**Nano-applications in Lighting**" was to investigate possible eco-efficiencies by means of the application of new nanotechnological products in the field of illumination. White LEDs were compared to conventional incandescent and compact fluorescent lamps; furthermore, the future potential of quantum dots was also investigated.

With light sources in use today, 97–99% of the life-cycle energy consumption occurs in the use phase. Materials consumption, in comparison, is of much lesser consequence. The crucial measure for the environmental assessment of light sources used as illumination, therefore, is energy consumption and associated emissions during the use phase. The current white LED, it turns out, compares favorably to the conventional incandescent lamp, but is at a disadvantage by a factor of three when compared to the compact fluorescent. Only with the further development of nanotechnology-based products with a significantly higher light efficiency, i.e., white LEDs with an efficiency above ca. 65 lm/W, will the LED become environmentally comparable to the compact fluorescent and useful in non-specialized lighting applications, i.e., for everyday lighting purposes.

The use of quantum dots as source of illumination will someday make possible even greater increases in energy efficiency. Quantum-dot technology is expected to find long-term application in the area of video displays, particularly in combination with OLEDs. However, the commercial application of quantum dots is still some years away.

The case study "**Risk Potentials of Nanotechnological Applications**" specifically focuses on the analysis and consideration of the potential risks of nanoparticles. In this study we explore the hazards that could arise due to the properties of the structures, substances, and materials utilized in the field of nanotechnology. We also give an account of the problematic effects for humans and the environment of selected nanoparticle structures and materials as already described in the scientific literature.

The behavior of nanoparticles is, in part, quite different from that of the same materials at the macro level; this is true even for identical compounds. Some extremely surprising effects and properties of nanoparticles and nanostructured surfaces can be found in many of the studies analyzed. They document a series of suspicious factors with respect to the possible toxic effects of nanomaterials and – particularly nanoparticles – on the environment and human health, all of which need to be seriously addressed. In addition to the essentially structureless nanoparticle, nano-structured materials such as nanotubes and buckyballs are of particular significance here.

However, the scientific results so far are preliminary in most cases and in part contradictory and they often consider only a small fraction of possible effects. The degree to which generalizations can be made or knowledge carried over from these individual studies is extremely questionable. The current state of knowledge is far too sketchy for a thorough risk assessment. Generalizations and toxicity classifications for nanoparticles appear to be possible in, at best, a medium-term time span.

Furthermore, the previous results relate principally to human-toxicological questions. We still know almost nothing about the ecotoxicity and behavior of nanoparticles in the environment. Much broader research is sorely needed. Taking into consideration production processes and the majority of applications, the issue of risks today, with respect to nanoparticles, does appear to be pressing, but of possibly limited scope if releases into the environment can be avoided. A point in favor of this is that many production processes occur in either an aqueous solution or in a closed system and, in many products, the particles or nanostructures are firmly bonded in a matrix. Nevertheless, over the course of the entire life cycle there still exist large gaps in our knowledge with respect to risk potentials.

1.3 Formative approaches to sustainable nanotechnology

Steering technological development by means of political intervention is either impossible or possible only to a very limited extent in complex modern societies. In spite of this, the course of such development is anything but chaotic. On the contrary, out of the interactions of the most varied agents a comparably stable course of development is often the result, one which can be accompanied in a formative way. The significance of independent paths of technological development over the course of time and the opportunities that these offer for the early identification of adverse effects on the environment and our health, and in-turn, for the timely assertion of influence, is depicted in the following illustration.

The following illustration makes clear that throughout the entire process, from basic and applied research through the development, use, and disposal phases, appropriate precautionary options can and must be developed; these options can then also be applied to further research and can also be influenced, as necessary through the use of Leitbilder. In each of the various phases, various players are (collectively) responsible. This can begin in the basic research phase on the basis of scientific paradigms whose results can subsequently lead to research and development efforts (research programs) in the area of applied research. It is our view that the most far-reaching opportunities to avoid potential environmental and health risks are to be found in these early phases of scientific and technological developments.

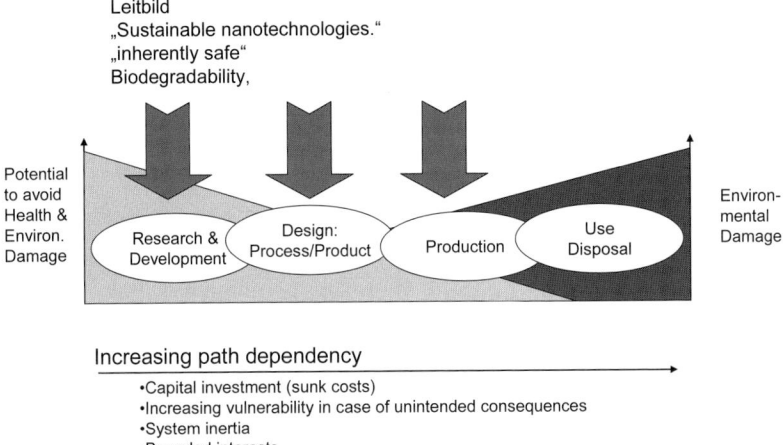

Fig. 2. Windows of opportunity for shaping technological development in the product life cycle[6]

The subsequent production process and product design phases are, in a sense, already predetermined inasmuch as they have as their foundation the imprint or character of the initial research and development phase. At the same time, there is a still relatively large degree of flexibility in process and product design that can be decisive for aspects such as "intrinsic safety" and/or possible environmental and health impacts in the subsequent phases. The options are more limited, but nevertheless still possible to a large extent.

In comparison, the number of options for shaping development in the production process and, thereafter, during use, and finally, disposal are much more limited. As a rule, at this point additional measures can still be implemented by means of the hazardous substance categories in product safety sheets, which regulate the handling of processes and products, or through the enforcement of disposal regulations.

In addition to the scientific and technical paradigms and development characteristics, the momentum acquired through the ever increasing lock-in of development pathways is also significant; this momentum arises, for example, from the participating investments involved and associated know-how and knowledge.

Scientific research on innovation and technological development has validated the importance of the Leitbild for the development of paradigms and technological trajectories (pathways) as well as for scientific revolu-

---

[6] Source: based on Rejeski et al. (2003)

tions and technological changes. Leitbilder motivate the formation of a group identity that serves to coordinate and synchronize the activities of individual players, reduce complexity, and structure perception. Among the most important prerequisites for their effectiveness, therefore, are vivid imagery and emotionality, their function as a guiding vision, as well as their relationship in equal proportion to vision and feasibility; in short, their degree of resonance in the minds of the players. Visual imagery plays an important role in clarification and the associated reduction of complexity. Above and beyond emotional value and content, the Leitbild motivates and provides direction. The Leitbild, therefore, can have a controlling or guiding effect and help to define the aims and direction of innovation. In this respect, it is possible to explore opportunities for steering innovation toward sustainable development with the help of Leitbilder. In the course of this project, three specific suggestions for sustainable nanotechnology Leitbilder, of varying scope, were developed: "Resource-efficient Nanotechnology," "Consistent and Intrinsically Safe Nanotechnology," as well as the long-term oriented Leitbild "Nanobionics."

In the various phases of the product life cycle (see Fig. 2), the various players that are involved in the design and whose negotiations are significant for the potential impact of the product or process each take their turn. Achievement of optimizations related to the product life cycle requires that communication channels exist between the various players along the entire value chain. With the exception of large firms that have comprehensive management systems or firms situated in adjacent production stages and capable of taking on the "system management" for the entire production chain, these communication channels are presently lacking in many areas. Communications channels along the value chain – and more so between the players and those possibly affected at the end of the product cycle – are often limited or not yet developed in the early phases of product development.

The most diverse institutions and circles of players are involved along the value chain; each alone has only limited opportunities for action, but through cooperation these opportunities can be significantly expanded. In principle, what one has is a typical governance problem in which the influences and possibilities for influence by the various players in the design and development process are not clearly perceived or assignable. As a result, in addition to the prognosis problem of the (prospective) technology assessment, with a view to the possible consequences of an emerging technology we also must consider the problem of complexity with respect to potential opportunities for influence. The possible consequences of innovations based on nanotechnology are, as already described, only predictable to a limited extent. The possibilities for steering or influencing each of the players or groups of players on an individual basis are, on the one hand, decisively affected by the individual dynamics of their social sub-systems, and on the other hand, limited by the developmental dependencies (lock-ins). It is important to create more flexible subsystems that react in a less compartmentalized fashion, but at the same time, it is important to exhaust the entire repertoire of possibilities for moving the market toward sustainability.

Therefore, in addition to the role of Leitbilder as potential instruments in technological development, further formative approaches and developmental instruments were outlined:

- The integration of safety, health, and environmental aspects in comprehensive quality management extending throughout the value chain.
- Sustainability-oriented nano-design in commercial research and development.
- Federal regulatory approaches

In view of the enormous prognosis problem with which the engineering results assessment has to struggle (even if the approaches arising out of the technology characterization have largely been exhausted), the importance of the concurrent approaches to specific development of nanotechnology or products and processes based on it must be emphasized. Along with those methods already introduced, related approaches such as constructive technology assessment (CTA) and real-time TA are also to be considered.

1.4 Conclusion and need for action

In the course of a three-stage approach, technology-characterization has made possible the identification of substantial hazards associated with some nanotechnologies – particularly in the case of nanoparticles and the possibility of a shift from self-organization toward self-replication – (hazard characterization), life cycle analysis has demonstrated potential opportunities for efficiency increases through selected applications of nanotechnology, and finally, basic approaches to sustainability-oriented design have been drafted. Among the significant conclusions that can be drawn from the study is, first of all, that potential risks from nanoparticles in non-contained applications already exist today and should not be ignored. Secondly, the results of the life cycle assessment demonstrate that the potential for significant environmental relief can be exploited, but that this does not fully apply to all areas of application and may entail calculated efforts. The technology characterization and the prospective (extrapolative) or concurrent life cycle assessment have also proven themselves as practical evaluation approaches, even for technologies, products, and processes still in development. Finally, it must be noted that Leitbild-oriented design – not in the least because of the findings in the first two approaches – will play an important role in further development of the technology toward sustainability.

At the operative level, in addition to the working with Leitbilder, the assessment, information, and communication instruments already mentioned here seem to be fundamentally well-suited for generating guidelines for action, particularly for small and mid-sized businesses (guidelines, development directives, management systems, etc.).

- Here, as a rule, the early phases of scientific and technological development offer the greatest opportunities for working toward sustainability; however they must be exploited. The development of formative Leitbilder offers itself as a valid approach.
- This should be augmented by concurrent and design-oriented processes, such as constructive technology assessment (CTA) or real-time technology assessment. Open communication processes throughout the entire value chain and product life cycle should be incorporated into scientific, business, and social organizations. Roadmaps for nanotechnological development represent another suitable instrument for integration and direction. Extrapolative life-cycle assessments – as presented here – should also be utilized and their results should be reflected back into the processes.
- As players in these processes, businesses have a substantial responsibility, which – particularly in newer EU environmental policy approaches such as REACh or in Integrated Product Policy (IPP) – have been repeatedly emphasized. The approaches mentioned assume at least a shared responsibility and in that respect take industry up on its promise. The development of integrated management concepts extending across the entire value chain, in which the aspects of health, safety, and environment (HSE) are recognized as quality assurance elements, would be a help. For the support specifically of small and medium-sized businesses, guidelines for nanotechnology-based sustainability-oriented design of products and processes would be tremendously useful in this context.

In addition to further development of such instruments, there is above all the ongoing need for further research. With respect to the assessment of nanotechnology hazards and risks, the need for research is particularly great with regard to:

- toxicological and ecotoxicological analyses within the framework of integrated research programs

- the systematization and classification of nanoparticles
- the behavior of nanoparticles and nanostructured surfaces in environment

Above and beyond the case studies investigated, which were very much focused on inorganic application contexts, there is a need for further research, particularly:

- with respect to still-to-be-completed or much further-reaching eco-efficiency potentials through the use of the principles of self-organization in the nanoscale dimension – in the inorganic as well as organic fields.
- with respect to a possible initial mid-to-long-term insidious transition from self-organization to self-reproduction in the areas of "active nano-systems" as well as particularly in the course of a possible coalescence of nano- with bio- or gene technologies.

It should furthermore be noted that future environmental assessment considerations of nanotechnological applications would be made significantly easier if material and energy flows (i.e., essential assessment data) for the relevant manufacturing methods were more easily accessible (or made accessible at all) and, if necessary, centrally collected.

If widely held expectations for nanotechnologies (and the innovations to be derived from them) are to be realized with respect to making significant contributions toward a sustainable economy, the approaches to a commensurate shaping of the processes, as described here and elsewhere, must be seriously considered. The potential benefits of nanotechnologies will not simply fall into our laps, but must be pursued with deliberation and effort. With this in mind, it is essential that technology, process, and product development be further accompanied by ongoing, concurrent assessment, with a view to precautionary risk management as well as to the effective utilization of the potential for sustainability benefits that are unquestionably associated with this line of technology.

2 Methodological approaches to the prospective assessment

In order to do justice to the complexity of the project's requirements, a three-step approach to prospective technology assessment and development of nanotechnology was followed.
- 1st Approach – prospective
  Technology characterization: Assessment of nanotechnology and its anticipatable effects by means of a characterization of the technology.
- 2nd Approach – process-concurrent
  Extrapolative life cycle assessment: Evaluation of sustainability effects utilizing life cycle assessment methods with specific current (already or about to be realized) applications in comparison to existing products and processes.
- 3rd Approach – formative
  Possibilities and limitations of the Leitbild approach: Leitbilder as a tool for directing technology development; accompanying processes and actor-specific concepts.

2.1 Characterization of technologies

2.1.1 Dealing with the unknown

Each and every engineering results assessment must struggle with the prognosis problem. How can problematic developments in technology be assessed; how do we record risks, secondary effects, and impacts that are entirely unknown? The prognosis problem in technological assessment has recently been further accentuated by the extensive attention being given to the problem of dealing with the unknown and the uncertain (cf. Wehling 2001; Böschen 2002; Wehling 2002). Generally speaking, there are two forms of the unknown:

1. The still-unknown. This is knowledge that, in principle, is acquirable, but not yet available, for example, because specific tests have not yet been conducted or specific know-how has not yet been acquired. There can be many reasons for this; it could be that a potential problem could not be anticipated because it had never occurred before and the impact model had not yet been fully worked out (the ozone-destructive impact of CFCs is a prominent example). All in all, it may well be the lack of resources (time, money, qualified personnel) that plays the central role in the still-unknown. A good example can be found in the almost 100,000 so-called "existing substances," which have not yet been tested for specific effects in accordance with chemical regulations.[7]
2. The unknowable. Here the limitations of knowledge lie not with the observer but rather in the realm of the object – that which is being observed. The reactions of unstable, complex, or dynamic systems to an intervention are basically unpredictable.[8] The reasons for this unknowability lie substantially in the architecture or instability of the system where the intervention is occurring. Of course, the intensity of the intervention (in quality and quantity) can (and, as a rule, will) further amplify this problem. Examples of extreme limitations on knowability include the oft-cited flutter of the butterfly's wings that in an extremely unstable weather system can set loose a tornado; unforeseeable reactions to gaps in food chains in ecological systems; and the unforeseeability of sporadic, spatially, and temporally isolated effects of climate changes (the Gulf Stream effect, for example).

Against this background, the prognosis problem in technological assessment becomes even more critical. How can one decide between and deal with these forms of the unknown? One possibility would be the selected application of a well-thought-out, systematic trial-and-error strategy, for which we will subsequently establish some initial approaches.

Even though it may be difficult for us as scientists to accept: Certainty is the exception! Uncertainty and incomplete knowledge is the rule. However, the difficulties in appropriately dealing with this situation are by no

---

[7] Approx. 3,000 of these existing substances are produced in Europe annually, with a volume of more than 1,000 tons per year. Only about 200 substances have been tested since 1981, when the law on chemical pollution came into force. This apparent failure is a major reason behind the re-alignment of European chemical pollution policy presently underway in accordance with the REACh approach, initiated by the EU white book on toxic chemicals. (cf. Europäische Kommission 2001).
[8] The ecosystem researcher Holling therefore refers in this case to "inherent unknowability"; cf. (Holling 1994)

means limited to scientists. For example, the constant passing back and forth of the burden of proof is, in light of the fundamental and wide-reaching borders of knowledge, a much-loved but entirely fruitless game. We can, of course, scientifically prove some potential dangers using already known impact models. But unless something can be determined experimentally or epidemiologically (or using model theory), it simply remains unknown. The harmlessness of substances or technologies cannot in any way be proven.

In light of the vast unknown and unknowable, we are essentially left once again with a process of trial and error – but with a well-thought-out, systematic strategy we can significantly improve on this. It would at first appear that fundamental pragmatics (resources), principles (unknowability), and likewise economics (international competition in innovation) would scarcely allow for any alternative to trial and error. However, given the circumstances of the unknown, the limitations of trial-and-error must be carefully considered. It is only justifiable in non-sensitive applications dealing with small, (theoretically) reversible steps. Even so, there are two significant considerations:

1. Step-by-step increment: In the course of the scientific and technological revolution, technologies have been developed on the basis of the mathematical and experimental sciences whose potential power and scale makes trial and error no longer possible. Due to the extreme potency of these technologies and the extremely long chain of events in space and time that they are capable of setting in action, each and every single error could have extreme, irreparable consequences.
2. Systematic approach and greatest possible containment: Even when the steps are carefully limited, the principle of trial and error should not be blindly and casually, but rather consequently and diligently implemented. Among the most important prerequisites for a methodical approach is a carefully worked out experimental setting (paying careful attention to previously acquired knowledge about possible and anticipatable impacts). Whenever possible, such experiments should, at first, only take place in demarcated, "experimental areas." Also essential is effective, targeted concurrent monitoring (to the extent possible), and last but not least, clear documentation of the experiments and results.

One could name several areas in which historical examples of the limitations of trial and error can be found; areas where the potentially adverse impact of a substance was recognized early on – either on account of its properties or the effects it produced – and yet the information brought no immediate consequences or the consequences came much too late with respect to the substance's handling.⁹ Such examples must certainly include the release of CFCs, with the global and irreversible consequences of the combined properties of this substance group (already known at the time): i) high degree of mobility, ii) little known (rare, foreign) substance properties, and iii) extreme persistence in the environment (see Table 4)[10]. One must also include the open deployment of asbestos (see Table 3) and the non-recallable release of genetically modified organisms capable of self-reproduction.

Table 3. Historical timeline of the asbestos controversy[11]

Year	Content of the controversy
1889	British factory inspector Lucy Deane warns of harmful and "evil" effects of asbestos.
1911	Animal testing offers "reasonable grounds" to support the suspicion that asbestos is harmful.
1911 and 1917	British Factory Department finds insufficient evidence to justify further measures.
1918	US insurers deny asbestos workers insurance because of concerns about illness-causing working conditions in the industry.
1930	The Merewether report in Great Britain finds asbestosis in 66% of all long-term workers in a factory in Rochdale.
1931	British asbestos regulations call for dust control in manufacturing only and compensation for asbestosis; but at the same time regulations are inadequately implemented.
1935–49	There are reports of cases of lung cancer in asbestos production workers.
1955	Richard Doll establishes lung cancer risk in asbestos workers in Rochdale.
1959–1960	Cases of mesothelioma found in workers and the general populace in South Africa.
1962/64	Mesothelioma cases in asbestos workers, people in the area surrounding the factories, and in relatives of factory workers found in Great Britain and in the United States, among other

---

⁹ Discussions of twelve case studies that correspond in part to the aspects mentioned here (asbestos, PCBs, TBT) can be found in the publication "Late Lessons from Early Warnings: the Precautionary Principle 1896–2000," published by the EEA.

[10] Due to this combination of properties alone, FCKWs were able to occur more or less "everywhere in the world," even in environmental conditions whose significance was still unknown (for a very instructive discussion of how forms of the unknown – and known – were dealt with, cf. Böschen 2000; Böschen 2002).

[11] Source: Gee & Greenberg (2001)

	countries.
1969	British asbestos regulations improve controls, but disregard end users and their cancer cases.
1982–1989	Asbestos controls for end users and manufacturers are strengthened; searches for substitutes are stepped-up.
1998–1999	EU and France ban all substantial uses of asbestos.
2000–2001	WTO upholds the EU/France asbestos ban against Canadian appeal.

Table 4. Historical timeline of the CFCs controversy[12]

Year	Content of the controversy
1907	Weigert conducts laboratory experiments proving that chlorine decomposes ozone.
1934	Similar experiments are conducted by Norrish and Neville.
1973	Global overview of CFCs by Lovelock et al. demonstrates their worldwide dissemination in the atmosphere.
1974	Moline and Rowland publish their theoretical arguments that CFCs could destroy the ozone layer.
1977	The USA bans CFCs in aerosols based upon "reasonable expectations" of detriment. Canada, Norway, and Sweden follow.
1977	World Plan of Action on the Ozone Layer is agreed to. The plan will be coordinated by UNEP.
1980	European decision limits the use of CFCs in aerosols, but this is offset by the increased use of CFCs in refrigerators.
1985	UNEP Vienna Convention is passed. It focuses on research, monitoring, information exchange, and, if necessary, restrictions.
1985	Farman, Gardiner, and Shaklin publish results indicating Antarctic "ozone hole."
1987	Montreal Protocol is signed; a phase-out with various timelines for ozone-damaging substances is established for developed as well as developing countries.
1990	Increased funding is made available to developing countries to lessen their dependence on ozone-harmful substances.
1997	Revision/enhancements are made to Montreal Protocol with goal of restoring chlorine levels by 2050–2060.
1999	Beijing amendment calls for efforts to stop illegal trade in ozone-damaging substances.

The examples suggest that ignoring signs (or not reacting to them) of potential dangers – which, as a rule, could already be seen at an early stage – has contributed to the unchecked proliferation of substantial environmental

---

[12] Source: Farman (2001)

and health hazards. The example of asbestos demonstrates the considerable consequential costs that businesses can be saddled with, for example, for production changeover (Eternit AG) and damage compensation claims (ABB) – which, in this case, could have endangered the ongoing existence of the firm.

It is therefore essential that clear limits be set on trial and error, particularly when the incremental step in innovation is very large and when applying extremely potent and deeply invasive[13] technologies with particularly wide-ranging anticipatable consequences (especially long reaction chains, for example, reaching across time and space); so that in the end, a single "non-recallable/non-stoppable" experiment (or massive deployment of less invasive technologies[14]) does not endanger an entire tract of land, ecological system, or the livelihood of future generations. For these reasons, specific limits must be placed on the trial-and-error approach, limits within which this approach can be responsibly implemented and, as far as that goes, be (democratically) allowed.

It comes down to this: No innovation is without risk. But how much risk? Unless we want to refrain from or prevent all forms of (technical) innovation, right from the outset, lack of knowledge about possible consequences is not a sufficient argument for refraining from particular forms of innovation. A "feeling of unease" or even the call for far-ranging measures along the lines of the precautionary principle, such as a moratorium, cannot be based merely on the "newness" of a technology and still-incomplete knowledge of possible problematic consequences. Novelty and a lack of experience are certainly good reasons for proceeding cautiously, as is always the case when one travels outside the routine of daily life; however, a justifiable "feeling of unease" – and thus far-ranging measures based on

---

[13] The concept of degree or depth of intervention builds on the considerations of Günther Anders and Hans Jonas (cf. Anders 1965); Jonas 1985): p. 42), and on the investigations into the scientific-historical and scientific-theoretical backgrounds to the development of experimental mathematical natural sciences, as well as the particularly interventionary technologies of atomic energy, synthetic chemistry, and genetic engineering (cf. Gleich 1989). The technology assessment criterion "degree of intervention" is explained in greater detail in Gleich 1998/1999).

[14] The latter was addressed in the "management rule" set out by the Special Commission on the Protection of Humanity and the Environment which says: "The tempo of anthropogenic inputs and interventions in the environment must be in a balanced proportion to the tempo of the natural processes relevant for the environment's responsivity (Enquete-Kommission des Deutschen Bundestages Schutz des Menschen und der Umwelt 1998): p. 46)

the precautionary principle – requires additional reasons.[15] These further justifications can be found either in the technology to be studied (for example, if a technology characterization strongly points to a high degree of intervention and an extremely high impact potential) or in the corresponding application context (for example, when intervening in the base functions of especially sensitive, instable, or important systems). One of the central challenges, therefore, of a methodically safe-guarded strategy of trial and error, is the problem of finding ways and methods to assess the extent of the unknown and to understand the possible consequences thereof.[16]

It becomes a matter then of the degree of the unknown and the level of potential danger (or risk). As a rule, these are determined by:
1. the nature of the intervention. Task and methods of the TA, here, is the characterization of the technology, for example, the identification of extremely potent high-risk technologies.
2. the degree of intervention. Task and method of the TA, here, is the identification of anticipatable cumulative effects, for example, with the help of a life cycle assessment.
3. the nature of the systems in which the intervention is occurring. The task, here, is the identification of extreme instabilities and the coping capacity and resilience of systems; one method for assessing the consequences of intervention in an ecosystem (in building a plant, for example) is the environmental impact assessment.

2.1.2 The "characterization of technologies" as an approach to prospective technology assessment

With a view to new, early-stage technologies in which neither the application contexts nor the degree of application are yet clear in most areas, the prospective TA (technology assessment) focuses on the technology charac-

---

[15] Cf. also Wiedemann et al. 2002), where starting points for the detection of corresponding "evident natures" are developed.
[16] "Dealing with the unknown" is not really new. Hazardous substances analysis (risk assessment) has long had internationally harmonized ways of dealing with the residual unknown factors. These include conventions for specifically extrapolating data (so-called safety factors, particularly in the carryover of results from animal experiments to humans or of short-term tests to possible long-term effects) and also in the utilization of so-called "default values." Toxicologists have been "reckoning" with the unknown in the truest sense of the word for some time now. But in other areas of the engineering results assessment, such methodical "reckoning with the unknown" is not yet well developed.

terization as a method of evaluation of the technology and appraisal of its anticipatable impacts. The technology characterization and the "nature" of the intervention (technologies, materials, processes, and products) already give us clear indications as to anticipatable impacts and also, in many cases, a basis for precaution-oriented development.

The engineering assessment, the knowledge of possible impacts from engineering, is, in a general sense, always dependent on knowledge of three fundamental elements (see Fig. 3):

1. an agent, i.e., the technology, the intervention, the material, etc., whose impact is being considered
2. an impact model, i.e., scientifically based conjectures as to how the agent affects a possible target (examples of such impact modules include the greenhouse effect; skin cancer as a consequence of stratospheric ozone destruction; and the carcinogenic, mutagenic, and reprotoxic effects (CMR) of materials)
3. a target, an endpoint or target system, upon which the agent acts (for example climate, ecosystem, organism, or organ)

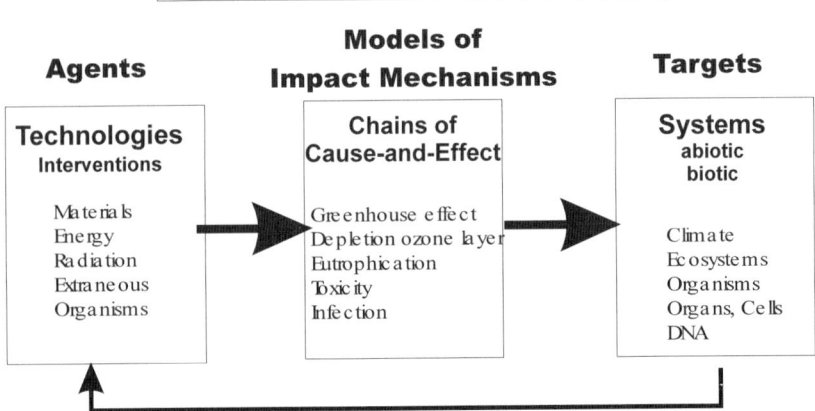

Fig. 3. A look at the technology assessment process: characterization of the agent[17]

---

[17] Source: Gleich (2003)

Any one of the three elements can be unknown. In the case of new technologies, the potential areas of application are, as a rule, to a great extent still unknown. Furthermore, new effects, such as can be expected with nanotechnology, mean that still unknown impact models must be anticipated. As to the technology itself – at least its basic principles – as a rule, one already has enough information, even in its early phases of development, to be able to assess potential hazards. With a view to possible impacts from nanotechnology, we must be prepared to deal with a particularly large number of unknowns, above all with respect to newer impact models and particularly with respect to the numerous impact locales of the targets of this fundamental innovation, which as one well can assume, will revolutionize entire industrial sectors and a large part of our life.

In accordance with the approaches proposed here for dealing with the unknown in the engineering results assessment, it becomes a matter of concentrating on that which, in this situation, is best suited for investigation: the agent. The focus of the TA should not be on the still uncertain target system and likewise still unknown impact models but instead focus more closely on the agent that will be applied to or act upon the target systems.

2.1.3 Technology characterization and types of hazards

The task and methods of the prospective TA, therefore, are the "characterization of the technology."[18] It is the question of which perhaps positive ef-

---

[18] The "technology characterization" approach utilized here as a prospective TA method is not any newer than the "reckoning with the unknown" mentioned in the previous footnote, i.e., dealing with safety factors. For the technology characterization as well, there are methodical parallels in the analysis of hazardous substances that have an already long established tradition. The "characterization of the (hazardous) properties of substances" (the so-called "hazard characterization" as an important step in risk assessment) concentrates on the problematic toxicological and ecotoxicological properties of substances. For risk management that is intended to implement the results of the risk assessment at the governmental and/or company level in preventative or safety measures, two types of substance properties have become especially relevant. These are, first, the CMR substances, carcinogenic, mutagenic, and reprotoxic properties (the hormone-like effects are increasingly included in the subcategory "reproduction toxicity"), the presence of which leads to direct management consequences – even before an exposure assessment; second, the properties that are particularly important for the likelihood of an exposure, such as volatility, solubility in water or fat, and particularly the ecological aspects: mobility, persistence (minimally degradable in the environment) and bioaccumulation (accumulation in

fects and potentials, but also dangers and risks, can be expected or already deduced from the characterization of nanotechnology. For this purpose, it is particularly helpful to look more closely at the qualities that make the new technology so interesting, i.e., the basis for its presumed potential for wide-ranging innovation or even a revolution in the way we live. There is good reason to believe that this technological potential and potency will not only have positive effects, but could also become a source of undesired secondary effects and problems resulting from its implementation. In particular:

1. The potency and degree of intervention of the technology. In assessing the potency and degree of intervention, one must, for example, consider the extent to which the corresponding technology affects the fundamental linkages and steering mechanisms in technical, biological, and environmental systems and then the effect of a corresponding potency and extension of space-time action chains. It is a question of the relationship of the technology and how it can be classified in comparison to other especially potent technologies and therefore possible "risk technologies." For especially potent technologies, one most likely must use the following technologies as a comparison in risk assessment and risk management: nuclear energy, with its catastrophic potential and the problems associated with the disposal of radioactive waste; the chemistry of hazardous substances, including the release of extremely mobile and persistent substances (for example, CFCs); and finally, the release of genetically modified organisms capable of survival.
2. The "new effects" attainable with the help of technology: In assessing these new possibilities, it becomes a matter of the extent to which the technology "only" improves or strengthens already known possibilities and effects and whether, or how, it truly can effect something new and never before seen. If these "new effects" are qualitatively not too intrusive, but at the same time open up a great many new fields of application, such that cumulative effects can be anticipated in the course of a most likely massive application, it is then advisable to instead use laser or computer technology and/or the so-called "new materials" for risk assessment and risk management.
3. The versatility of the technology with regards to possible effects as well as possible areas of application. The assessment touches on the question as to the extent to which one is dealing with a "key technology" and/or

---

the food chain). According to the plans of the European Commission (Europäische Kommission 2001; Europäische Kommission 2003), in the future *very persistent* and *very bioaccumulative* substances (vPvB) will require an approval process – even without a substantiating impact model.

"fundamental innovation."[19] Depending on the extent to which this might be the case, the technology is most likely comparable to information and communications technology or modern biotechnology.

2.2 Environmental profile assessments

Building on this characterization of nanotechnology, the concurrent ongoing assessment approach to the prospective TA is followed by an ascertainment of sustainability effects using specific application examples in comparison to existing products and processes, whereby, in accordance with the life cycle assessment method, the focus is on the substance and energy flows and their associated ecological opportunities and risks. The involvement of current fields of application means that the time frame of this approach is accordingly much, much smaller than is the case with the technology characterization. In such life cycle assessment approaches, only those developments that are almost ready for market can be assessed. And comparisons can only be made in those areas of application of the new technology that compete with already existing technologies, processes, or products (and for which the corresponding data is already available).

Since the appraisal of potential environmental impacts is the top priority, the environmental profile assessments for the in-depth case studies are modeled in their approach on the life cycle assessment methodology. However, they are unable to meet the high standards required, since the data available for new products and processes – and, to an extent, also for those products and processes used in the comparisons – is far from complete. The life cycle assessment is the most extensively developed and standardized methodology for assessing the environmental aspects and product-specific potential environmental impacts of a product. Many potential applications of nanotechnology will, on the basis of fundamental deliberations (and aspects of the technology characterization), be associated with desires for the realization of enormous resource efficiency improvements. The potential validation of these desires, that is, the compari-

---

[19] With all the hope for an economic upswing, growth and prosperity, that is generally associated with such a characterization as an enabling technology or fundamental innovation, whereby, of course, even in the case of an extensive economic structural change, not only winners but also losers must be expected. The latter will be particularly found in those very sectors and enterprises that are not capable of replacing their "old" technologies and solutions with nanotechnical innovations.

son of the eco-efficiency potentials of nanotechnological solutions, products, and processes to existing applications, is possible with the help of the life cycle assessment.

In accordance with EN ISO 14040, a life cycle assessment consists of the following steps:
- Definition of the goal and scope of the investigation
- Life cycle inventory analysis
- Life cycle impact assessment
- Life cycle interpretation

The following illustration makes clearer the relationship between these steps.

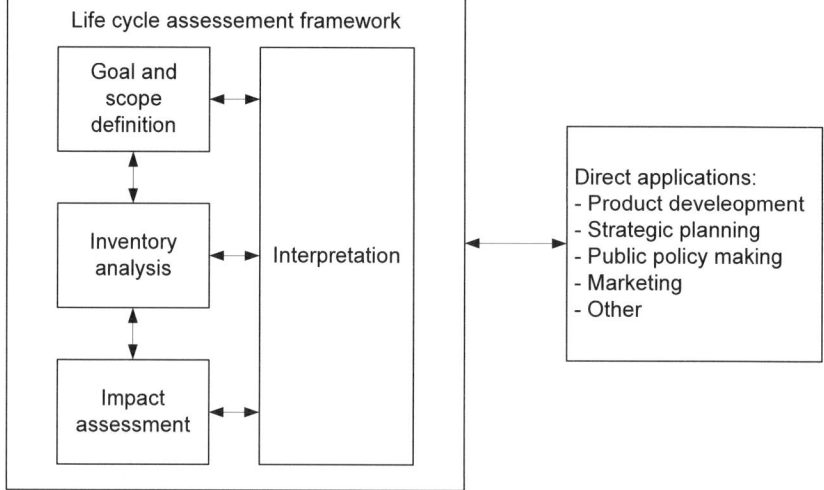

Fig. 4. Steps in the preparation of a life cycle assessment[20]

The directional arrows between the individual LCA steps should make clear the iterative nature of the process, i.e., the results of further steps are always fed back into the process, possibly resulting in further changes and iterations.

In the first step, the goal and the scope of the study are defined. The second step, the life cycle inventory analysis, involves the collection, compilation, and calculation of data. The life cycle assessment, as its name suggests, generally looks at a product or service over its entire life cycle. Specific substance and energy data therefore must be collected for each life cycle stage. Essential data includes, on the input side, the consumption

---

[20] Source: DIN EN ISO 14040 (1997)

of raw and ancillary materials, including energy inflows, and on the output side, product data, including air and water emissions and waste data.

In the life cycle impact assessment the data of the life cycle inventory analysis is organized (classification) and summarized (characterization) according to its environmental relevance. In this way, the resource drawdowns and emissions that occur over the course of the product life cycle are brought into the context of environmental impacts in debate among experts and the public. The following table lists some examples of some impact categories and the substances contributing to them.[21]

Table 5. Impact categories and contributing substances[22]

Impact category	Contributing substances and factors
Demand on resources	Consumption of renewable and non-renewable resources (crude oil, natural gas, coal, minerals, lumber, etc.)
Greenhouse effect	Carbon dioxide (CO_2), methane (CH_4), nitrous oxide (N_2O), among others
Stratospheric ozone depletion	chlorofluorocarbons (CFCs), brominated and halogenated hydrocarbons, among others
Human toxicity	Volatile organic hydrocarbons (VOCs), organic solvents, airborne particulates, benzene, heavy metal compounds (arsenic, cadmium, mercury, lead, nickel, etc.) Sulfur dioxide (SO_2), nitrogen oxides (NO_x), fluorides, hydrogen fluoride, hydrogen chloride, carbon monoxide (CO), among others
Ecotoxicity	Sulfur dioxide (SO_2), nitrogen oxides (NO_x), fluorides, hydrogen fluoride, hydrogen chloride, lead (Pb), cadmium (Cd), copper (Cu), mercury (Hg), zinc (Zn), chromium (Cr), nickel (Ni), adsorbable organic halogens (AOX), among others
Summer smog	Nitrogen oxides (NO_x), methane (CH_4), volatile organic hydrocarbons (VOCs), among others
Acidification	Sulfur dioxide (SO_2), nitrogen oxides (NO_x), ammonia, (NH_3), hydrochloric acid (HCl), hydrogen fluoride (HF), among others
Aquatic eutrophication	Nitrate (NO_3^-), ammonium (NH_4^+), chemical oxygen demand (COD), total phosphorous, total nitrogen, among others
Terrestrial eutrophica-	Nitrogen oxides (NO_x), ammonia (NH_3), among others

---

[21] The toxicities and the impact on natural area resources (in part also the use of these resources) are not being presently being taken into consideration in most life cycle assessments because of the problems of standardization and quantization.

[22] Source: Ankele & Steinfeldt (2002)

tion	
Impact on natural area resources	Extraction of raw materials (for example, mining of coal and ore), demands for areas of a certain ecological quality (for example, agriculture)

The last step of a life cycle assessment is the analysis and interpretation. This includes the derivation of conclusions and concrete recommendations for action for the planned application or utilization.

The life cycle assessment, as already mentioned, is one of the most thoroughly developed and standardized of the TA methods. Nonetheless, like all methodologies, its also has its weaknesses, blind spots, and deficiencies. Some of these shortcomings need to be mentioned.

1. There are impact categories for which generally accepted impact models do not yet exist. This is particularly true in the relevant categories of human and environmental toxicity. For example, with regard to scale, a consideration of the impact of fine dust particles (the PM-10 risk deals with the potential toxicity of particles smaller than 10 µm) in life cycle impact assessments is therefore already doomed to failure because of its reference to weight and not particle size in nanotechnology applications. Other "qualitative" impacts that cannot be directly correlated with the levels of material and energy flows, such as structural impacts on ecosystems, also cannot adequately be captured.
2. Furthermore, in life cycle assessments, neither the technical risks nor the potency of applications are considered.

In our view, the life cycle assessment alone cannot generate a comprehensive evaluation of the environmental impacts (environmental compatibility) of products and processes (see also Klöpfer et al 2007). In addition to the life cycle assessment, a comprehensive environmental assessment plan for nanotechnology applications should – at a minimum – take the following further assessment methodologies into consideration:

- Risk analysis
- Hazard characterization, particularly with the help of technology characterization including analysis of the degree of intervention or intrusion
- (Eco)toxicological analysis
- Environmental impact assessment (particularly with a view to plant facilities)

Using an actuarial approach to risk (risk = occurrence probability x extent of damage), the risk analysis is dependent upon statistical information (alternatively, models). In risk analysis, too, there arise almost insurmountable difficulties with respect to technologies or applications still in development. Statistical data is usually not available. In this situation, it is necessary to fall back on a hazard or risk analysis (hazard characterization), as was done in this project, in the form of a general "characterization of nanotechnology."

The analysis of the degree of intervention is part of the technology characterization and looks at the extent to which steering mechanisms or elementary connections are impacted by the possibly extreme potency of a product or process and its potential for expanding space-time impact chains. The concept of "degree of intervention" is intended to establish the nature of a technology, to reflect, for example, the fundamental qualitative differences between the splitting of stone and splitting the atom (cf. von Gleich & Rubik 1996).

The toxicological analysis includes tests for acute toxicity; chronic toxicity; corrosiveness; skin and eye irritation; carcinogenicity, mutagenicity, and reproductive toxicity; skin sensitization (allergic reactions); weakening of the immune system, etc.

Such a set of methods, whose corresponding prioritization must be adapted to the specific application context, should make possible, on the one hand, the assessment of the environmental impacts and risks associated with specific applications as compared to existing applications, and, on the other hand, the analysis of accessible eco-efficiency potentials. However, because information and data for specific application contexts is often lacking and available working resources are limited, the application of this complex method set has significant limitations.

In the project, these methodological shortcomings were minimized through the establishment of priorities in the selection of the specific application contexts. From the entire spectrum of nanotechnological applications, four case studies with anticipatable eco-efficiency potentials were selected on the basis of a preliminary examination and qualitative preliminary assessment. Possible risks and potential dangers of nanotechnology applications – with an emphasis on the toxicity risks associated with nanoparticles – were furthermore analyzed and discussed, and human toxicological indications with respect to potentially adverse effects of nanoparticles and nanostructured surfaces were recorded. The basis for this were, first, investigations into the impact of ultra-fine particles (UFP) on health, and second, investigations on nanoparticles, but also fullerenes and nanotubes, which to date have mostly been carried out in experiments with animals. Investigations of possible ecotoxicological effects are still almost non-existent. In addition to the main steps of the analysis, these toxicological case studies essentially represent an overview of the scientific literature and expert opinions in this field. The discussion on TiO2 nanoparticles is taken as an example.

2.3 Approaches to shaping technological development

The examination of opportunities and risks in terms of potentials and problems in sustainability is part of a process that deals with innovations and their direction, their further advancement, realization, and obstruction. A sufficiently refined understanding of the processes of technological development and innovation is crucial if one is to successfully influence the development and direction of new technologies.[23] Innovation generally takes place within complex structures, so-called innovation systems,[24] which can include many players; these can be differentiated by type as well as by level of communication and action and can exist at the market sector, company, regional, national, and international level. If attempts at shaping development are to be successful, it is important to be aware of not only the type of system and the level at which it operates, but also the time profile of innovation processes.[25] Not so long ago, the story of technology was seen as an ongoing process of ever more advanced development, a more or less autonomous evolutionary process. Such linear phases of development do, in fact, occur. The development of technology can occasionally follow and evolve under its own, relatively large, momentum. It proceeds then – and that would be only one of several attempts at explanation – in small steps of improvement, along a technological trajectory. The direction of development is therefore largely determined by this path or trajectory. Changes in direction and alternatives here are especially difficult to achieve. Further-reaching opportunities to influence development would be limited in this model to the phases (time frames) before the pathway is entered or to those moments of upheaval shortly before a pending change of direction. Before the start of a development pathway, i.e., before the re-

---

[23] Illuminating these processes of emergence (genesis) and development of technologies and innovations is the subject of research into the genesis of technology (cf., for example, Dierkes 1997) and innovation research (cf., for example, (Sauer & Lang 1999; Hübner 2002).

[24] Cf. (Nelson 1993; Freeman 1995); in research on the genesis of technology one also refers to technical genesis networks.

[25] As a part of the innovation research network Framework for Innovations Leading to a Sustainable Economy (RIW – cf. www.riw-network.de and Hemmelskamp 2001), funded by the German Ministry of Education and Research, the project SUBCHEM ("Options for Facilitating Innovation Systems for the Successful Substitution of Hazardous Substances") is involved in attempts at standardization against a background of case studies. The meaning and usefulness of windows of time is being investigated by the project SUSTIME (Innovation, Time, and Sustainability – Time-Frame Strategies in Environmental Innovations Policy).

sulting technological "lock-in," decisions on direction of development are constantly being made at the most diverse possible junction points. Here, a large degree of freedom exists. Once on the developmental pathway, however, this becomes much more difficult. Possibilities to change course only first arise once more when problems occur for which the development pathway can provide no solution or when a competitive development trajectory is seen to be emerging.[26] Such windows of opportunity do not only occur in the course of technological momentum. In the case of capital investments, in particular, the investment cycle of a firm, for example, also plays an important role.

The players involved in these innovation systems each have and use their own specific opportunities for influence. The range of possible influence varies, but is, as a rule, strictly limited. The probability that a player (or group of players) in the system would be capable of directing the course of the entire innovation process is likewise low. The likelihood of directing such innovation systems "externally," may well be even less. More likely, still: In light of the complexity of technological genesis and innovation processes, interventions aspiring to a "directing" function are not only more or less futile, they tend to be dysfunctional or even contraproductive. This is not intended to question the much cited "primacy of politics" in technological development and therefore in a field that very much – and ever more and more – influences living conditions in the modern industrial society. Nor is it an apologia for technical momentum or an issue of technocracy or expertocracy. Politics and democratic (participatory) influence should, however, not focus on "directing," but rather – and this is the current practice – on influencing and shaping.

This study takes as its basis a further concept for influencing and shaping design that takes into account not only the direct possibilities for influence, but also the mediative and indirect possibilities. The study thus focuses on varying aspects of potential for influence, specifically, informative processes and approaches, for example, the possibilities offered by Leitbilder, the preparation of road maps, and discursive processes such as constructive technology assessment (CTA, cf. Rip et al. 1995). In

---

[26] The "model" that is supposed to make the development of innovations and technologies understandable (cf. particularly Dosi 1982; Dosi 1988) is very reminiscent of the model with which Thomas Kuhn attempted to explain the development of science as derived from paradigms: a) The development of paradigms, then b) "normal science" within the realm of the paradigm, then c) accumulation of unsolvable problems within the realm of the paradigm, and finally d) entry into the phase of the paradigm shift, the "revolutionary science" (Kuhn 1975).

particular, an emphasis is placed on those tools that improve orientation and dissemination of information in and among firms, for example, eco-design, development directives, etc. Government regulatory approaches that could be relevant for nanotechnology are also outlined. As an example, reference is made to the new toxic chemical regulations (REACh – and their focus on the precautionary principle). And finally, we look at definitions and compliance with regulatory frameworks and so-called safety barriers, within whose boundaries research and innovation processes are able to more or less freely operate. These safety barriers should, however, also allow the optimal utilization of the numerous small potentials for the maneuvering of the players within the innovation system, particularly for businesses, who will increasingly be confronted by the potential consequences of their actions.[27] In the process, a particular emphasis is placed on the new environmental policy approaches such as REACh and the integrated product policy approach, both of which extend along the entire value chain. They increasingly assume a shared responsibility and force businesses, at least in principle, to more diligently fulfill their responsibilities.

Newer approaches to operationalization of the precautionary principle

A life without risks is not possible. The conversion to sustainable industry is dependent on innovation. Innovation and risk are inseparably connected. The path itself to sustainable industry is full of risks. The fact that a technology, process, or substance is new is not in and of itself a sufficient reason for implementing far-ranging measures based on the precautionary principle. For this, greater reasons for concern are necessary. In many instances, we will have no other choice, even in the future, than to simply test or "try out" certain things. But the trial-and-error method, upon which we so often must fall back, has its limits, as already described above, and is appropriate only for small – largely reversible – steps.

When it can justifiably be assumed from the very beginning that particular projects, technologies, and operations will have global and irreversible consequences, the trial-and-error method would be irresponsible. The introduction of CFCs was such a case. The precautionary principle comes

---

[27] As an example one could look at the problems of ABB with regards to demands for damage compensation due to the deployment of asbestos, but also the activities of the insurers and reinsurers, whose sensibilities have greatly increased in this respect. For example, reference could be made to the nanotechnology study of the Swiss Re (Hett 2004).

into play when dealing with enormous impact potentials, i.e., extremely large qualitative developmental steps in the case of individual innovations (highly pervasive) or large quantities or quantitative rates of increase in the case of numerous small steps (cumulative effects).

When introducing innovations, "caution" can be defined as the main guideline for the implementation of the prevention principle, Specifically, caution with respect to the developmental increment and degree of intervention (here, the technology characterization provides the essential information) as well as with respect to quantities deployed and the rate of introduction (for which production quantities and their increase rates are essential information).

In the draft for new EU hazardous chemicals regulation (REACh – Registration, Evaluation and Authorization of Chemicals), both approaches to operationalization of the precaution principle can be found. Thus the quantities utilized in the introduction of an innovation (substance and/or application) determine to a large extent the level of risk analysis effort accordingly necessary. The level of chemical testing and the scope of the data accordingly required are determined by the corresponding production quantities. The most extensive amount of data is required for larger quantities and volumes; for smaller quantities, significantly less; and for the amounts commonly used in research, almost none.

On the other hand, the qualitative aspects of an innovation with respect to pervasiveness or intervention play a wide-ranging role in this new approach to hazardous chemicals regulation. Specific chemical properties, such as carcinogenicity, mutagenicity, and reproduction toxicity (CMR), already determine to a large extent the risk management prerequisites – even before a more precise exposure assessment – an at least as important element of any risk analysis. A further, particularly interesting example for the implementation of the precautionary principle is the proposed handling of very persistent, very bioaccumulative (VPVB) chemical substances. Because these properties alone make a high-level and irreversible environmental exposure very likely (and if the property of mobility is added, very quickly a global exposure), these materials require government approval, even when there are no indications as to concrete toxicological or ecotoxicological factors. Here the risk management measures follow directly from the characterization of the substance (or technology), without the existence of a scientific impact model (toxicological or ecotoxicological suspicion) or even a completed risk assessment.

Leitbilder as tools for shaping technological development

With a view to the players in research and in development in the research institutions and industry, there arises the necessity for "gentler" forms of influence, suited to the complexity of innovation processes and the limited autonomy of social subsystems. Among the most important elements of such gentler methods of influence – and this, too, as already mentioned, is a result of research on the genesis of technology – is without a doubt the Leitbild, i.e., a form of "guiding principles" mission statement. If one could succeed in working out the (more or less explicit) Leitbilder that underlie the development of nanotechnology and then influence their further interpretation and modifications, one would succeed in opening up the field just a bit more to public and democratic discourse.

Successful technical innovations are, as a rule, based on linking the doable with the desirable, i.e., combining (often new) technical possibilities with as-yet-unsatisfied present or future social needs and unsolved problems. Exactly this is one of the features and most important functions of Leitbilder, that these two aspects, the desirable and the doable, are linked together.[28] Therefore: Leitbilder influence the development of technology, and with the help of Leitbilder, the course of this development can in some cases be selectively influenced. There is almost no question as to the fundamental effectiveness of Leitbilder, even when empirical evidence as to their specific effect in a particular line of technology is scarce and possibly difficult to come by.[29] Whether and to what extent Leitbilder can be deliberately employed as an instrument of influence or shaper of technologies is still largely unclear (cf. for example Mambrey et al. 1995; Meyer-Krahmer 1997; Kowol 1998). A theoretical answer to this question may be difficult – one would have to look at practical attempts. In strategic corporate management, one often finds reports on a comparatively successful deployment of various "management by" approaches, including "management by guiding principles," in other words management with the help of

---

[28] Cf. Dierkes and others 1992, where, incidentally, reference is also made to a further interesting parallel with respect to the difference between the "doable" and the "desirable" in the Leitbild discussion. It is a question of the difference between and the typecasting of innovations into "technology push" and "demand pull" innovations. With both types, i.e., in the further development of technical possibilities, as well as in the development and formulation of societal needs and problems to be solved, *Leitbilder* play an important role.

[29] Cf. Hellige 1996); *Leitbilder* may be particularly of importance for so-called fundamental innovations – in the phase before technological lock-in of a development path and in times of upheaval before a technological change of direction.

corporate Leitbilder (cf. Matje 1996; Blättel-Mink & Renn 1997; KPMG 1999; Bea & Haas 2001).

To exert influence and shape development with the help of Leitbilder, one must understand the prerequisites for effectiveness and modes of action of successful Leitbilder. Leitbilder motivate the formation of a group identity that serves to coordinate and synchronize the activities of individual players, reduce complexity, and structure perceptions. Among the most important prerequisites for an effective Leitbild are vivid imagery and emotionality; functionality as a guiding vision, as well as a balanced approach to the desirable and the doable; in short, its ability to resonate in the minds of the players.[30] The vivid imagery is important for the clarification and therefore reduction of complexity. Above and beyond emotion and values, the Leitbild motivates and provides orientation. The reference to feasibility is important in avoiding too unrealistic, utopian visions. For an effective Leitbild, therefore, the degree of abstraction should not be too great. Starting points for application and operationalization should be clear and direct. "Sustainable development" as a Leitbild could be too complex, too abstract, and too defensive. In the main discourse, at least, as it was begun in Rio on the topics of environment and equity, climate protection, species conservation, and the protection of natural resources, the sustainability goal is far too concerned with the bare survival (availability of resources, carrying capacity) and too little with the "good life." The focus, therefore, was rather more on what is necessary and less on what is desirable. A sector-oriented Leitbild, such as "Sustainable Construction and Housing," the Leitbild outlined and initially operationalized by the Special Commission on the Protection of Humanity and the Environment could be more effective (The Special Commission of the German Bundestag on the Protection of Humanity and the Environment 1997). Key concepts, such as resource efficiency, sufficiency (frugality), and consistency (embedment of the social metabolism in the natural), much discussed at the strategic level, are also very likely to be too abstract. An example of an effective technology-oriented Leitbild, neither too abstract nor too literal and at a mid-level of operationalization, is undoubtedly the recycling concept. Bionics (nature, the example) and, in the English-language world, "green chemistry" have also proven effective.[31]

---

[30] One finds numerous interesting overlaps in the areas of "image richness," "reduction of complexity," "structuring perception," "motivation," "development of group identity," coordination and synchronization," "connection to reality," and "preferred tools and idealized ways to reach a solution" between the *Leitbild* concept and Thomas Kuhn's paradigm concept (Kuhn 1975).

[31] (Cf. Ahrens & Gleich 2002)

Most assume that risk management is based upon an as-complete-as-possible risk assessment, i.e., upon wide-ranging knowledge of effects. But precaution and risk avoidance also are a part of risk management. In the case of insufficient knowledge, both strategies must turn to possible hazards and impacts models. Leitbild-oriented design is proposed in this case as an additional contribution to a strategy of precaution and risk avoidance. We will always know too little about the possible consequences of technologies and the degree of intervention into social and natural systems. All the more so, when we attempt to anticipate the consequences of technologies that have not yet been fully developed or even introduced. One approach to a solution for appropriately dealing with the unknown consists, first, in an attempt at a prospective engineering results assessment (by means of technology characterization and concrete context assessment), and secondly, in an attempt at leitbild-oriented technological development. That means that we not only approach the principally unsolvable prognosis problem of the engineering results assessment with the assumption that the approaches and tools of a prospective engineering results assessment will be further developed, but rather that we also will attempt to implement the "prognosticated," or more precisely "the desirable," right from the very beginning. Finally, "active realization" may well be, as before, the most promising method in order to bring real development into alignment with prognosis.

It remains clear, however, that even when combining the two approaches, no absolute guarantee of safety can be achieved. The much-cited zero risk is not going to happen. The active pursuit of the goal of risk minimization should yield better results than the previously all-too-often-followed scenario, in which this goal first became a priority only after the child had fallen in the well.

3 Technology-specific impacts of nanotechnology

3.1 Characterization of nanotechnology

Dealing with the new eventualities that nanotechnology makes possible is a genuinely complicated affair. On the one hand, the scientific, technical, and economic revolution it offers, particularly its truly fundamental newness is justifiably emphasized. On the other hand, and here the enthusiasm diminishes, such technical revolutions hardly occur without problems, side effects, and unintended consequences; with groundbreaking, new technological opportunities, as a rule, new risks are also introduced. It is a fact, objects in the nanodimension behave in part quite differently than they do in the macroscopic world. Among these "new behaviors," for example of nanoparticles, are properties relevant to toxicity and ecotoxicology such as solubility, reactivity, selectivity, and catalysis, as well as propagation and mobility in the environment and in organisms, for example permeability of cell membranes and penetration of the brain via the nose and olfactory nerve. That means that we cannot assume that what we know from our experience in the macro world will hold true or have an equivalent in the nano-world.[32] Unfortunately, our risk assessment mechanisms and procedures have so far insufficiently made adjustment for these "new qualities." Facilities for the manufacture of nanoparticles are still approved based upon the classical regulations and worker safety and environmental risk

---

[32] It turns out, for example, that titanium dioxide, in wide use as a white pigment and considered to be an "inert" substance in the macro- and micro-world, i.e. chemically and biologically non-reactive, can be chemically quite active in its nanoparticle form (see Jefferson & Tilley 1999). It remains to be mentioned that nanoparticular TiO_2 has already been in use for some time now in many of the sun protection creams with a particularly high sun protection factor. And also, that in experiments on animals the toxic and inflammatory effects of ultra-fine TiO_2 could be substantially reduced by means of a "hydrophobic coating" (see Höhr & Steinfartz 2001).

assessment procedures of the macro world, where the weight of the emitted substances plays the central role and not the number of particles.[33]

It is worthwhile then looking at the qualities that make nanotechnology so interesting but presumably will also be the source of unwanted side effects and consequences, particularly: i) nanotechnology's potency and pervasiveness, ii) the "new effects" made possible with the technology's help, and iii) its versatility.

Let us begin with the simplest building blocks. The behavior of "nanoparticles" is generally quite different from that of larger particles of the same composition. The drastically greater proportion of particle surface to volume, for example, generally leads to a different chemical reactivity and selectivity. In the nanodimension, such surface-related effects can vary tremendously, both qualitatively and quantitatively (new and/or stronger effects). At the same time, quantum effects – for which there is no counterpart in the macroworld – gain in significance. The technical practicality of quantum effects are, however, dependent on a "well-defined" environment; in the grubby "real world" they become disordered or displaced and tend to founder. Unexpected environmental impacts caused by quantum effects are therefore rather unlikely. Conversely, one must be aware of the effect of real-world "impurities" and "disturbances" on the well-defined system of quantum effects; these could become relevant sources of errors and "technical failures," possibly leading to far-reaching consequences (a question of intrinsic safety and fault tolerance in the application system).

The prevailing "nano-vision," a vision strongly influenced by macroscopic and "mechanistic" thought, seems to be one of a controlled and systematic structure of matter or nano-objects, one built, so to speak, block by block, atom by atom, molecule by molecule. Much more interesting and fascinating, however, are the "self-organization processes" that take place at the nano level, in which molecules systematically order and align themselves, as if guided by an unseen hand. Should we someday succeed in technically rallying these self-organization processes, enormous savings in materials, energy, but also otherwise unachievable "external control tasks" might become possible. On the other hand, such self-organization processes could possibly overflow and become uncontrollable.

There are still more points that are important for a technological characterization of nanotechnology. (For now we will stay with nanoparticles.)

---

[33] See in this regard the legal arguments submitted in the approval process for a plant facility of the Bayer subsidiary H. C. Starck in Laufenburg am Hochrhein; for further considerations that include (then) not yet feasible nanotechnological developments such as nanorobots, see (Forrest 1999) (Reynolds 2001) (Haum et al. 2004).

Nanoparticles are, as a rule, so small and light that when airborne they remain floating in the air rather than sinking to the ground. Due to their small size, the particles are extremely mobile and can be carried great distances by air and water (Lecoanet & Wiesner 2004; Dionysiou 2004). They can also travel through cell membranes in organisms and possibly even the blood-brain barrier; direct passage to the brain via the olfactory nerve can also occur (Oberdörster et al. 2002; 2004). Nor are they easily controlled – or directly perceivable. Scientifically their presence and therefore more importantly their "responsibility" for certain effects is not easily demonstrable. On the other hand, suitable countermeasures against the uncontrolled proliferation of nanoparticles may well make it possible to set clear limits. One measure in their favor is that by means of adhesion and cohesion processes they generally rapidly attach themselves to other objects and readily bind together to form large agglomerates.[34] In this way – as a contribution to their own inherent safety, so to speak – their "nanoproperties" are quickly nullified.

Not the least important issue in the course of a technology or product life cycle assessment is a close look at how the nanoparticle and nanostructured materials are produced. The question arises, as to whether the production efforts needed to achieve a required "chemical purity" of a perhaps extremely seldom or specific chemical element or precise particle size become excessive and whether this effort economically and environmentally pays off in the long run, when one looks at the entire product life cycle. It is important to observe recycling procedures, particularly with respect to the quality of the substances utilized. Are toxic or ecotoxic substances or substances with as-yet unknown effects being employed in these areas? Are rare natural substances being "mobilized" and set free in an area, so that one must deal with problematic effects on the metabolic activities of organisms or the ecosystem?

Based on these fundamental characteristics of selected branches of nanotechnology ("passive nanosystems," such as nanoparticles and their associated smallness, mobility, altered reactivity and selectivity, catalytic effects, solubility, quantum effects; "active nanosystems" and self-organization effects), possible beneficial and/or anticipatible problematic effects could now – with additional attention to further aspects such as production expense and substance quality – be derived in a next step (see Table 6).

---

[34] These tendencies of nanoparticles also make the technical handling much more difficult, a phenomenon that – as the technical problem of so-called "stickiness" – is already known from microsystem technology.

The use of self-organization principles in "active nanosystems" is not to be confused with absolute technical control over the "building blocks," whereby matter is assembled by means of technically controlled intervention – so to speak, atom by atom. The chances of such perfect control of nature as compared to the use of self-organization principles may be comparatively small. The mechanistic building block principle is generally dependent on specific conditions that are technically only realizable with great effort (clean room technology?) and therefore likely already embedded in an "intrinsically safe environment."

In table six the preliminary results of a technology characterization of selected nanotechnologies is presented. Prospects and risks are derived from the characteristics of the technology. This makes possible statements about anticipatible consequences at a point in time when the application contexts are not yet known. It also provides findings that can be incorporated into debates about effective formative models. Using formative models to influence research, development and innovation processes helps us more accurately strive for the desirable and at the same time avoid the undesirable.

Table 6. Nano-qualities and their anticipatible positive and negative impacts and effects[35]

Nano-quality		+ Positive impacts and benefits / – Problems and potential dangers	Nano-example	Assessment approaches
A. Passive nano-systems			Nanoparticles and nano-structured surfaces	
Smallness and mobility of particle	+	Selective use for resource- and environmentally efficient technology	Openly handled manufactured nanoparticles*	Life cycle assessment. dispersal and exposure models. (eco-)toxicological testing, animal testing, epidemiology
	–	Dust-inducing, mobile via the air pathway, non-settling, remain airborne. Absorbed by the lungs and alveoli. Penetrates cell membranes, in some cases also the blood-brain barrier. Mobility. Persistence and solubility as indicators for bioaccumulation and environment hazard		
Defined particle/layer sizes and purities	+	Selective use for resource- and environmentally efficient technology		Life cycle assessment. entropy balance. question of "ecological amortization".
	–	Increased production costs, material and energy streams, and use of resources		Toxicology, ecotoxicology, relationship between "natural" and "anthropogenic" material streams
Material qualities	+	Possible replacement for substances dangerous to health and environment		
	–	Health and environmental dangers due to hazardous (rare) elements or substance groups in environmentally open applications		

---

[35] Source: Modified in accordance with Gleich (2004) and Steinfeldt (2003)

			Examples	Research needs
Adhesion, cohesion, agglomeration	+	"Intrinsic safety" due to adhesive, cohesive, and agglomerative tendencies of nanoparticles		Dispersal and exposure models. (eco-)toxicological testing, animal testing, epidemiology, atmospheric chemistry, risk analysis
	-	Behavior of nanoparticles or fibers "set free" in the environment. Mobilization and inclusion effect of nanoparticles on toxins and heavy metals (piggybacking)		
New chemical effects, altered behavior: solubility reactivity selectivity catalysis phase transitions	+	Use of the altered behavior for resource- and environmentally efficient technology, for example Use of the catalytic effects for more efficient chemical processes or in the area of environment	Nanotubes as catalysts in styrene prodution* Autocatalysts Nanocoatings* New composites	Life cycle assessment, dispersal and exposure models, (eco-)toxicological testing, e.g. also for allergies / sensitization. Animal testing, epidemiology, atmospheric chemistry, Risk analysis
	-	The large specific surface of nanoparticles leads to in part entirely unexpected changes in: solubility (important for absorption in organisms and propagation in environmental compartments), reactivity (unexpected technical, chemical [explosiveness] and toxic effects), selectivity (unexpected toxic effects), altered and/or increased catalytic effects (unexpected chemical and toxic effects), photocatalytic effects (atmospheric chemistry?), temperature dependence on phase transitions (unexpected chemical, technical, and ecotoxicological effects)		

3.1 Characterization of nanotechnology

Category	+/-	Description	Examples	Methods
New physical effects, modified optical, electrical, magnetic behavior	+	The relatively small number of atoms in nanoparticles cancels out the quasi-continuous state and leads to varying optical, electrical and magnetic properties. Selective utilization of effects and modified properties for resource and environmentally efficient technology, e.g. the GMR effect, Tyndell effect, quantum effects, tunnel effect	CMR sensors Nanotubes in FED diplays* OLED* White LEDs*	Life cycle assessment. For technical systems: FMEA - Fault-tree analysis
	−	Generally dependent on high-purity, highly defined "technical environments." Surprises (in the case of non-compliance) can be expected (technical failure) In the open environment, side effects – of the quantum effect, for example – in organisms and eco-systems are less likely		
B. Active Nano-systems			From biomimetic material synthesis to nano-bio units and assemblers	
Self-organization	+	Selective use for resource / environmentally efficient and consistent technology	Self-Assembled Monolayers. Template-controlled crystallization	Risk analysis. Analysis of the degree of intervention or intrusiveness.
	−	Problem of insidious transition from self-organization to self-replication in nano-biotechnology Danger of uncontrolled development	Future vision: self-replicating and breeding nano-bio units and nanobots	Life cycle assessment. Environmental impact analysis. Scenario method

* These examples will be more closely investigated in the case studies (see chapter five).

3.2 Description and evaluation of production methods

In addition to this characterization of nanotechnologies at the technological level, we will now examine in a second step the relevant production methods for nanoparticles and nano-structured surfaces in more detail.

Due to the interdisciplinary nature of nanotechnology, an enormous wealth of methods for the production of nano-scale products can be found in the literature. Products can, for example, be differentiated according to their nano-scale basic structure: particle-like structures (e.g. nanocrystals, nanoparticles, and molecules), linear structures (e.g. nanotubes, nanowires, and nanotrenches), layer structures (nanolayers), and other structures such as nanopores, etc. Materials can also be produced from the gas phase, the liquid phase, or from solids in such a way that they are nano-scalar in at least one dimension. An overview of selected nanotechnology-based products and their application status is available in the appendix (Fehler! Verweisquelle konnte nicht gefunden werden.).

When evaluating the manifold nanotechnological products that are currently being applied or that are still in the development stage with respect to their method of production, from our point of view most products can be classified into one of the following seven basic process technologies:
- Gas-phase deposition
- Flame-assisted deposition
- Sol-gel process
- Precipitation
- Molecular molding
- Lithography
- Self-organization

3.2.1 Gas-phase deposition (CVD, PVD)

Gas-phase deposition is an important class of processes, which are mainly used for the deposition of thin films from gaseous source materials; they are also used for the precipitation of monodisperse powders, quantum dots, etc. Gas-phase deposition can roughly be divided into chemical (CVD = chemical-vapor deposition) and physical (PVD = physical-vapor deposition) gas phase deposition. In all the methods the source materials from the gas phase are transported to the substrate surface where they form a thin film (Köhler 2001).

In the case of PVD a solid source material placed into a vacuum is transferred into the gas phase by physical means (e.g. thermal energy). The particles precipitate and form a thin film on a substrate. The different methods of PVD can be distinguished by the means of evaporation, i.e. the method of heating. In thermal evaporation, the conventional method, the starting material is heated in high vacuum in a crucible until it evaporates. Further methods include sputtering, arc-evaporation, molecular beam epitaxy (MBE) and ion plating (Fraunhofer IPA/IST & BAM 2004b). The general process of PVD is shown in figure 5.

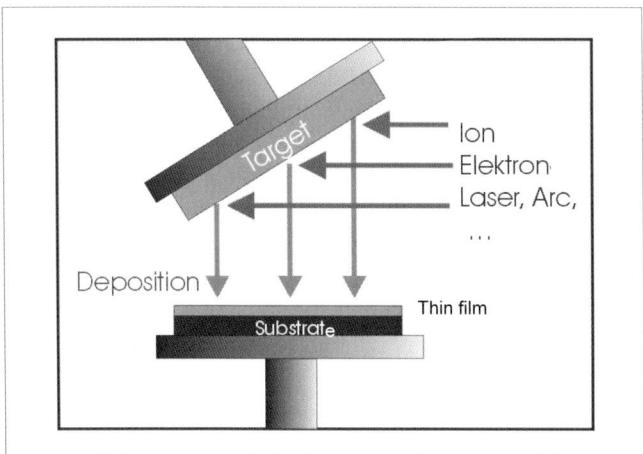

Fig. 5. The PVD method[36]

CVD includes all methods that utilize the chemical reaction of a gaseous source material on or close to a substrate to precipitate a solid product. The gaseous source material is fed into a reaction chamber where it is then thermally decomposed. The energy is supplied either thermally, by excitation of the reactants with a plasma, or by electromagnetic radiation. A portion of the formed intermediate product is adsorbed at the substrate, where a heterogeneous reaction leads to the formation of a film. Those volatile reaction components that do not precipitate on the substrate are removed via the vacuum system (Fraunhofer IPA/IST & BAM 2004a). Important CVD methods include: Thermal CVD, plasma-activated CVD (PACVD), photo-assisted CVD and catalytic CVD, which is increasingly used in the production of carbon nanotubes.

---

[36] Source: Fraunhofer IPA/IST & BAM (2004b)

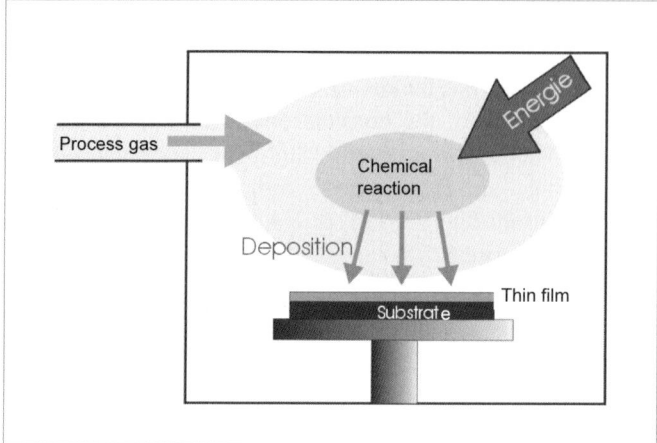

Fig. 6. The CVD method[37]

3.2.2 Flame-Assisted Deposition

In the various flame-assisted deposition methods, nanoparticles are formed through the decomposition of liquid or gaseous source materials in a flame (Maichin 2002). Among the best known methods are flame-spray pyrolysis, flame hydrolysis (the Aerosil method), and flame synthesis. Hydrogen or hydrocarbons serve as fuel. One substantial advantage of this method can be seen in the fact that the flame already supplies the necessary energy. That way fine-grained powders can be produced without the need for extensive pre-treatment or post-treatment. Furthermore, the production of oxides can be realized without the need for costly vacuum equipment or reactors.

By varying the concentration of the reactants, the flame temperature, and the dwell time of the source material in the flame, particle size and crystal structure can be controlled. By adjusting the process parameters, the particle size can be precisely defined. These methods currently enjoy wide use in the industry. Their use for the production of nano-structured particles has been standard in the chemical industry for years. Industrial soot (carbon black), for instance, is produced by this process. In the flame, at a temperature far exceeding 1000°C, carbon black particles of only a few nanometers in size are formed by rapid cooling with water. They im-

---

[37] Source: Fraunhofer IPA/IST & BAM (2004a)

mediately combine to form larger aggregates. They are utilized in this form, for example, in automobile tires.

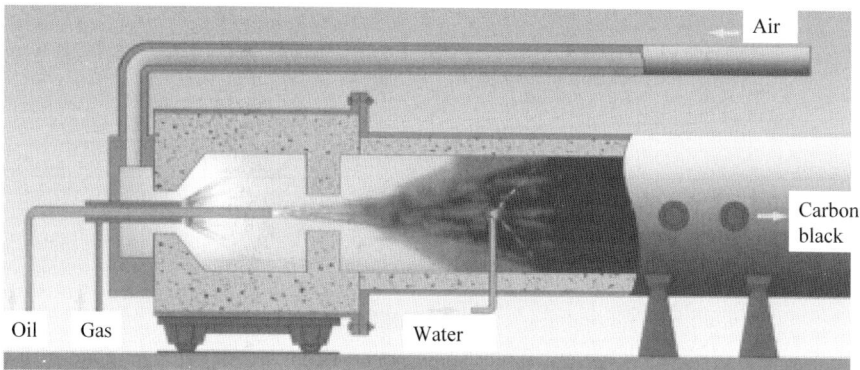

Fig. 7. Illustration of a reactor for the production of Carbon Black®[38]

3.2.3 Sol-gel process

The sol-gel process is an exceptionally important wet-chemical method for the production of a variety of nano-technological products such as powders, thin coating, aero gels, and fibers (Fig. 4).

---

[38] Source: Kühner (1999)

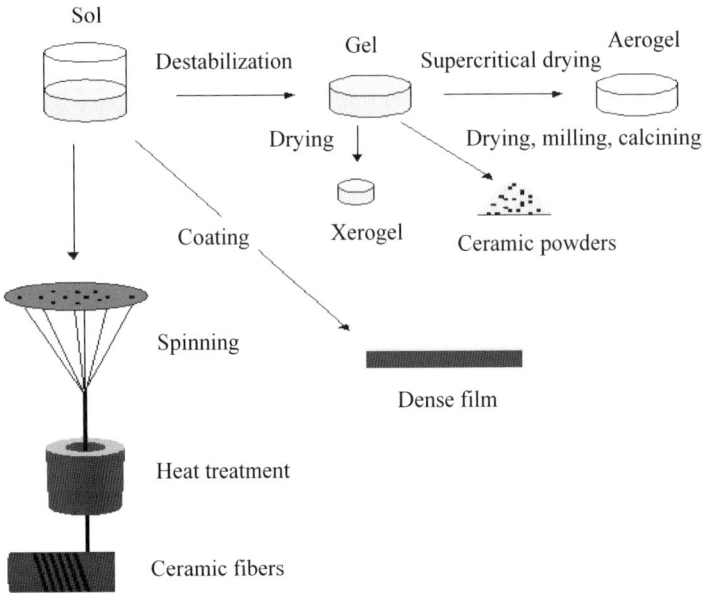

Fig. 8. Application examples for sol-gel products[39]

In the first step, nano-scale colloids or nanoparticles are formed by the reaction of the liquid components. Subsequently, the sol is converted into the gel state by one of two main methods: the molecules primarily formed in the sol can either continue to grow and aggregate by chemical reaction until they form a single macromolecule filling all available space or particulate sols can coagulate until a gel is formed, which is then stabilized by electrostatic repulsion. By destabilization of the sols or gels, respectively, nanoparticles of defined sizes can be precipitated. One of the most promising possibilities offered by the sol-gel process is the combination of organic and inorganic components into tailor-made organically modified products. Another advantage is the simplicity of the method: production can be carried out in a test tube (IMST 2002).

---

[39] Source: IMST (2002)

3.2.4 Precipitation

Precipitation is a chemical method whereby nanoparticles are deposited from solution. In this way, for example, metallic nanoparticles can be produced. The material to be deposited is dissolved in a solvent, usually water. The addition of appropriate reagents initiates the precipitation. Either the composition of the solvent is modified in such a way that the substance to be deposited then becomes less soluble or insoluble, or a new compound is formed that has a significantly lower solubility than the concentration in solution. The formation of the nanoparticles proceeds step-by-step from crystalline seeds or amorphous primary particles to particulate agglomerates. It is crucial that the seed formation rate or nucleation rate, respectively, be faster than the growth rate of the particles. In a continuous precipitation process, the particle-size distribution as well as the structure of the agglomerates can be fine-tuned by process engineering, i.e. by choosing both the appropriate flow conditions and particle-particle interactions (Schwarzer 2001).

3.2.5 Molecular molding

In this method, a highly cross-linked polymer is synthesized in the presence of a template molecule. A template, here, is a molecule whose specific geometry controls growth, structure, and configuration of the system being built up. The functional groups of the monomer bind to those of the template and lead to a replication of its outer shape. Subsequently, the template molecules are removed by extraction. Vacancies with binding sites having a defined spatial arrangement are left in the polymeric network thus formed. The guest molecule that is to be selected is structurally similar or identical to the template and can be molecularly recognized and bound. Using this method it is possible, for example, to isolate individual molecules from a mixture (IGVT 2003).

52   3 Technology-specific impacts of nanotechnology

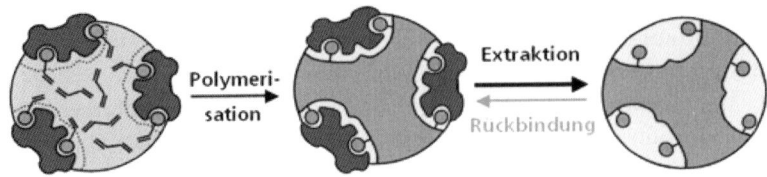

Fig. 9. Molecular molding[40]

3.2.6 Lithography

The lithographic methods used for the production of nanostructures can be grouped into two classes: In parallel methods, the entire surface is simultaneously structured. In the serial method, the structure is successively inscribed (Cerrina & Marrian 1996). Among the first group of methods are optical lithography, electron beam and ion beam projection lithography, atom lithography, and x-ray lithography. Serial methods include electron beam and ion beam writing as well as scanning probe lithography.

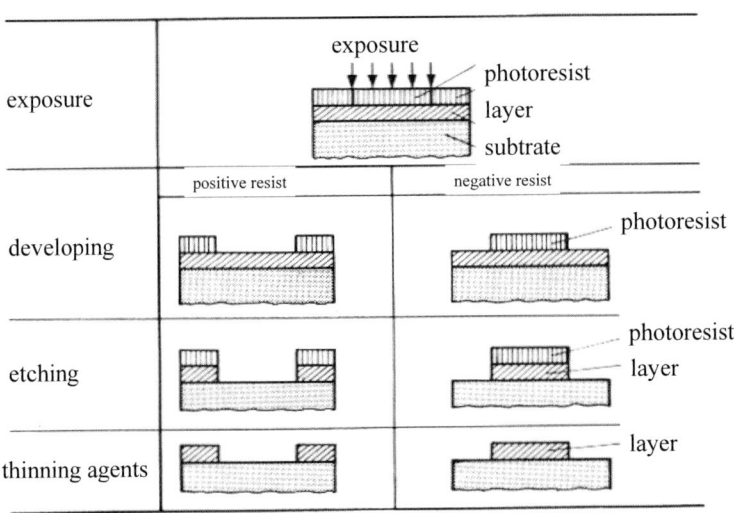

Fig. 10. Schematic depiction of the lithography process and the subsequent etching process[41]

---

[40] Source: IGVT (2003)
[41] Source: Rossi (2000)

Optical lithography is the most frequently used method for the production of nanostructures. The semiconductor structures created with this method ultimately comprise the foundation of the entire electronics industry. In optical lithography, light or x-rays are projected through patterned masks onto a sample surface coated with photoresist. After development of the photoresist, the reproduced pattern is usually transferred onto the substrate by etching. The pattern resolution, i.e. smallest reproducible detail, depends on the wave length of the light used.

In electron beam lithography, a focused beam can be used to write directly on the substrate (electron beam writing) or the pattern can be defined by a mask (electron beam projection). In electron beam writing a narrow electron beam with a diameter in the nanometer range scans the substrate surface, which has been coated with photoresist. The structure is written into the photoresist and can then be transferred to the substrate by etching. In the electron beam projection method, the electrons are deflected by a transparent silicon nitride foil and thus directed onto the substrate.

Ion beam lithography works in a very similar way to electron beam lithography; however, here direct patterning of a work piece without photoresist and etching is also possible.

3.2.7 Self-organization

Self-organization cannot be regarded as a technological method in the proper sense. In fact, it is a molecular building scheme by which growth- and structure-forming processes in nature also take pace. In this process, individual building blocks such as molecules, atoms, and particles combine to form functioning units. The organization process is regulated by interactions between the individual building blocks (Eickenbusch et al. 2003). A pivotal characteristic of self-organization is that the information for the structural formation is stored in the individual building blocks. The most important role here is played by geometric shapes and charge distributions that permit the building blocks to fit together only in specific ways.

One method for the production of self-organized structures that is already being used in products intended for the marketplace is the formation of self-assembled monolayers.

Self-Assembled Monolayers (SAM): Long-chain organic molecules form densely packed single-layer structures by adsorption on oxidic and metallic surfaces. In this way, ultra thin films can be made with a structure that is pre-determined by the configuration of the substrate surface atoms, which form chemical bonds with the adsorbed molecules (see Fig. 11). These layers are referred to as self-organized monolayers (Jelinski 1999).

54   3 Technology-specific impacts of nanotechnology

If the molecules deposited in the monolayer contain, in addition to the functional group needed for the bond to the substrate, an additional one at the opposite end of the molecule chain, this can be used as a template for the selective deposition of inorganic materials.

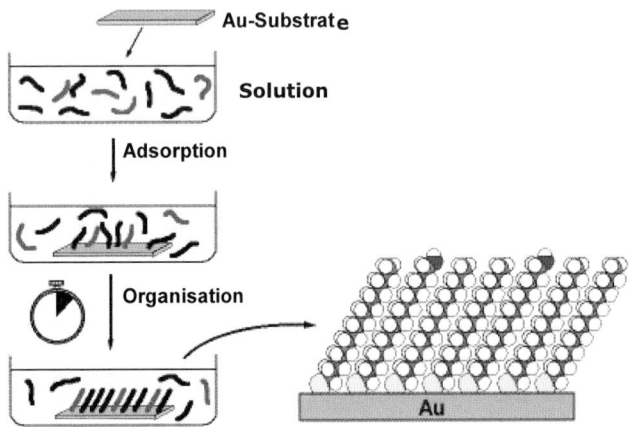

Fig. 11. Production of self-organized monolayers[42]

3.2.8 Qualitative assessment of the various production methods

In the following table, these process technologies are evaluated qualitatively with regard to the technological effort required and, accordingly, energy inputs, as well as the risk potential for the release of nano-structural materials.

---

[42] Source: Engquist (1996)

Table 7. Qualitative assessment of production methods for nanoparticles and nano-structured surfaces

Source material	Method	Facility requirements	Energy requirement	Material throughput	Risk of release of nanoparticle emissions	Product Risk of release of nanoparticle emissions
Gas-phase Deposition CVD PVD	Gaseous Medium	Technologically very demanding method: vacuum chamber facilities; high-quality clean room;	Very high due to high fixed-energy requirements and high energy requirement depending on evaporation temperature of source material	Quantity throughput: small Material throughput: close to 100% Yield: medium	With respect to workplace: minimal due to technology (vacuum) and contained equipment With respect to the environment: low to medium, because gaseous starting materials can be treated in waste-air purification facilities	Low, if nanomaterials are tightly bound in a fixed layer Medium, if nanomaterials are bound in a layer tightly sealed in the product Large, if nanomaterials are produced as loose nanoparticles or tubes, respectively
Flame-assisted deposition	Gaseous	Routine, no pre- or post-treatment	Medium, only energy requirement is for flame (temp. ca. 1200–2200°C)	Quantity throughput: very high Material throughput: very high Yield: high	With respect to workplace: medium, direct emissions possible due to equipment leaks and during bottling of nanoparticles With respect to the environment: low to medium, because gaseous by-products can be treated selectively in waste-air purification facilities	Large, because nanomaterials are produced as loose nanoparticles

Sol-gel process	Liquid or dissolved	Routine chemical engineering	Small	Quantity throughput: medium to high Material throughput: medium to high; dependent on the reaction parameters Yield: medium to high, unreacted compounds possible	With respect to workplace: small to medium, because process takes place in a liquid medium; air emissions depend on temperature and vapor pressure of the process medium With respect to the environment: small to medium, because although emission of nanomaterial possible via contaminated process media and waste water, it can be selectively treated in waste-water treatment facilities	Low, if intermediate products are liquid or nanomaterials are tightly bound in a fixed film in the end product. Medium, if nanomaterials in the end product are bound in a film but not stable long-term
Precipitation	Liquid or dissolved	Routine chemical engineering	Small	Quantity throughput: medium to high Material throughput: medium to high; dependent on the reaction parameters Yield: medium to high, unreacted compounds possible	With respect to workplace: low to medium, because process takes place in a liquid medium; air emissions depend on temperature and vapor pressure of the process medium With respect to the environment: low to medium, although emission of nanomaterial via contaminated process media and waste water possible, it can be treated selectively in waste-water treatment facilities	Low, if nanomaterials are tightly bound in a fixed film in the end product Medium, if nanomaterials in the end product are bound in a film but not stable long-term

Molecular molding	Liquid	Routine chemical engineering	Small	Quantity throughput: medium to high Material throughput: medium to high; dependent on the reaction parameters Yield: medium to high	Related to the work place: low to medium, because process takes place in liquid medium, air emissions depend on temperature and vapor pressure of the process medium With respect to the environment: low to medium, because emission of nanomaterial possible via contaminated process media and waste water, but those can be treated selectively in waste-water treatment facilities	Small, because the product contains only nanostructures but no loose nanoparticles. However, nano-structured surfaces can also have various unexpected effects.
Nanolithography	Solids	Technologically very demanding method; extensive washing and etching procedures; high-quality clean room:	High, due to high fixed-energy demand (clean room)	Quantity throughput: medium to low Yield: medium	With respect to workplace: low to medium, because process takes place in liquid medium. Air emissions depend on temperature and vapor pressure of the process medium With respect to the environment: low to medium, because emission of nanomaterial possible via contaminated process media and waste water, but those can be treated selectively in waste-water treatment facilities	Low, because the product contains only nanostructures but no loose nanoparticles

				Quantity		
Self-Assembled Monolayers	Liquid or dissolved	Routine chemical engineering	Small	Quantity throughput: medium to high. Material throughput: medium to high: dependent on the reaction parameters. Yield: medium to high	With respect to workplace: low to medium, because process takes place in a liquid medium: air emissions depend on temperature and vapor pressure of the process medium. With respect to the environment: low to medium, because emission of nanomaterial possible via contaminated process media and waste water, but those can be treated selectively in waste-water treatment facilities	Low, if nanomaterials are firmly bound in a fixed layer. Medium, if nanomaterials are bound in a layer tightly sealed in the product. Medium, if nanomaterials in the end product are bound in a film but not stable long-term

In contrast to other processes, these all possess a higher potential for the release of nanoparticle emissions in a gaseous form, which could lead to direct emissions in the workplace and the production of lose nanoparticles, for example in flame-assisted deposition. With the other methods, the risk potentials can be regarded as low, provided that the emissions can be treated by appropriate air-purification and waste-water treatment systems. However, it must also be noted that not only nanoparticles but also nano-structured surfaces can lead to a variety of reactions (e.g. catalytic effects).

Furthermore, some of the process technologies presented here, already in use as fundamental technologies in the microelectronics and optoelectronics industries, require an enormous technological and energy effort, while yielding a rather small absolute quantitative output. The technological degree of intrusiveness, risk potential, controllability, and material and energy demands of the production methods for nano-structured materials discussed so far are comparable to many established methods in the chemical industry, computer technology, and in the area of so-called "new materials"; therefore risk assessment and risk management efforts can draw on the experience gained in these existing fields.

3.3 A digression: "nano-visions" and their risk assessment

From the very beginning of nanotechnology, far-reaching visions of the future have been intrinsically associated with its development and history. One must assume that these futuristic visions – and not least the numerous and ubiquitous science fiction images of "nano-engines," consisting of only a few atomic layers, and "nanorobots," moving about in the bloodstream and fighting cancer cells, bacteria, and viruses – were instrumental in the very establishment of the idea of a "nanotechnology." It was precisely these visions that finally led to the breakthrough of the umbrella concept "nanotechnology" as a paradigm. They were able to arouse the emotions needed to carry the concept into politics, news, and public debate and to mobilize enormous amounts of funding. And of course authors such as Drexler played an important role in this.[43]

---

[43] Regarding this, Summer (n.d.) writes: "The rapid development of nanotechnology is to a large extent originally due to the visions of future technological possibilities and their implications for society by a few scientists, among them especially K. Eric Drexler (see Drexler et al. 1986; Drexler 1991; Drexler 1992). Following the example of Drexler such a multitude of mainly American scientists and science journalists have created futuristic scenarios about nanotech-

Where futuristic visions such as these are to be found, reactionary horror scenarios are not far behind. An increasingly critical debate on the dangers of proliferating nano-devices began, one in which old fears were mobilized, from the magic broom run amok to robots capable of overpowering mankind (see Drexler 1986, Joy 2000). The issues of "grey goo" and mechanical self-reproduction regularly reoccur as horror scenarios intended to evoke the dangers of nanotechnology.

The starting point for such scenarios was also provided by Drexler, who developed the idea of "assemblers," machines that would be capable of producing any kind of object by extracting atoms from the environment and assembling them into a new object, atom by atom.[44] These assemblers could be programmed and equipped with their own power supply. Since they would be able to produce anything they could accordingly reproduce themselves, which, in the case of malfunction could lead to an alteration of the entire biosphere (grey goo). This vision led to the demand for specific safety measures in future nanotechnological development.

Currently there is a strong trend to try to end these debates by questioning the technological feasibility of such "assemblers," mainly by means of arguments based on physics and science.[45] However, (current) technological infeasibility is not an especially strong argument in the light of the technological developments of the past 150 years. At the same time major debates are being conducted and important reports published on the "convergence of information technology, nanotechnology, biotechnology, genetic engineering, and cognitive science to improve the progress of mankind" (Rocco & Bainbridge 2002). The following two illustrations are from this report on "converging technologies" and illustrate the associated visions of convergence and technological architecture of the 21st century.

---

nology that from this a downright new literature genre has developed, which is apparently also used by research politics in order to give a popular illustration of nanotechnology." p. 9 foll.

[44] Drexler follows the line of Feynman's famous talk in 1959 "Plenty of Room at the Bottom," which today is viewed by many as nanotechnology's founding myth. Feynman 1959. http://www.its.caltech.edu/~feynman/plenty.html

[45] Often problems of power supply, thermodynamics and (self-)control are treated; see, e.g., the debate between Drexler and Smalley in Chemical and Engineering News December 2003
http://pubs.acs.org/cen/coverstory/8148/8148counterpoint.html.

3.3 A digression: "nano-visions" and their risk assessment

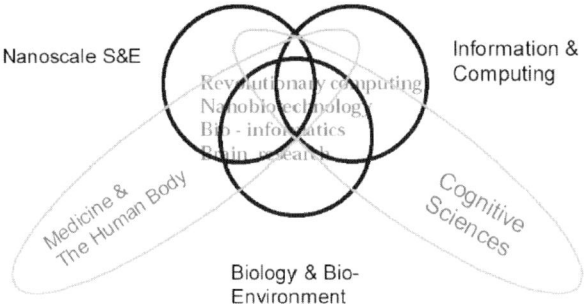

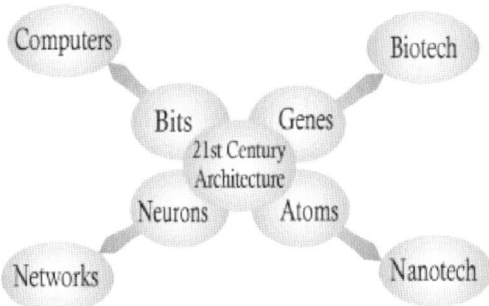

Fig. 12. Visions of convergence and a technological architecture of the 21st century[46]

Here at least one thing should be clear: Debate about such long-term visions of fascinating possibilities and the associated horror scenarios will continue and will scarcely be ended by decree. The social debate on nanotechnology has in large part thrived on these.

However, this does not mean that such populist visions and speculations must decide scientific technology assessment – on the contrary. We should concentrate not on speculation but on that knowledge already available to us today. This means that with regard to "converging technologies" as described above, a risk assessment based on a "characterization of the technology" must be further and resolutely pursued. On that basis some impor-

---

[46] Source: Rocco & Bainbridge (2002)

tant differentiations can already be made, which can also be found in the recently published report of the Royal Commission, particularly the distinction – which we have already made several times – between passive and active nanostructures, and furthermore the distinction between a "mechanical" and a "biological" path of development for nano-devices. With respect to the assessment of the risk of uncontrolled behavior or the transition from self-organization to self-replication, there are large differences between these two paths. The probability that "mechanically structured" nanorobots (nanobots or self-assemblers) would (inadvertently) self-replicate, probably is quite minimal. This is not equally true for nanostructured units based on "biological" molecules – DNA in particular, which is currently the subject of many nano-biotechnological experiments and projects – or on organelles or even cells. To a certain extent, even the simplest SAMs, the self-assembled monolayers based on "polar" molecules, are model systems after the example of "biological membranes." So the biological path to nano-devices should be, by far, the more promising (and more worrisome). David Pescovitz of UC Berkeley's College of Engineering sums it up accordingly: "Biology is the nanotechnology that works" (Pescovitz 2004). Or as the Royal Society puts it in its report "Nanoscience and nanotechnologies – opportunities and uncertainties": "Where we can find self-replicating machines is in the world of biology" (The Royal Society 2004).

The difference between self-organization and self-replication is important – and long before the point, as here in this project, at which hopes are being placed in steps toward a sustainable economy based on nanotechnologies using the principles of self-organization of molecules.[47] Technologies based on principles of self-organization promise – already at the inorganic level (for example in the case of template-controlled formation of structures, directional crystallization, and the already mentioned SAMs) – reductions of material and energy needs as well as supervision and control expenses. Self-replication refers to the self-reproduction of which organisms are capable – including, for example, genetically altered organisms. Self-organization and self-replication are therefore very different "capabilities." Nonetheless, transitional states between them are entirely

---

[47] And these are quite widespread expectations. See, e.g., the technology analysis of nanobiotechnology (VDI-Technologiezentrum 2002, S. 105f) The bottom-up synthesis of material structures, making technological use of self-organization phenomena, was singled out as particularly promising and a further technology assessment especially for this topic was recommended; with good reason in our opinion. In fact, this appeared in 2003 (see VDI-TZ 2003) and expressly underscored this potential.

3.3 A digression: "nano-visions" and their risk assessment

possible. In the course of biological evolution, the capability of self-replication must, in fact, have developed – at least once – out of the self-organization function, of course, over an immeasurable period of time.

In a hazard or risk assessment with a view to advanced applications of self-organization (e.g. a combination, conversion, or merger of nanotechnology and biotechnology, and perhaps also nanotechnology and robotics), the aspect of self-reproduction has a certain relevance that simply cannot be dismissed on grounds of technological (in)feasibility. The move from self-organization to self-reproduction certainly would bring with it new dimensions of risk.

As long as nanotechnology limits itself to the handling of inorganic molecules, such a step, from the self-organization of molecules to self-reproduction and multiplication of nano-devices – unless deliberately attempted – should be quite unlikely. However, in the case of a merger of nanotechnology with genetic manipulation of organisms capable of self-reproduction, such a step is entirely viable.

4 Assessment of sustainability effects in the context of specific applications

Carrying on from this characterization of nanotechnology and its current manufacturing processes, the process-concurrent assessment approach follows and investigates the sustainability effects of specific applications as compared to existing products and processes, with a focus is on the ecological opportunities and risks. The goal is preparation of environmental profiles in the course of in-depth analysis of selected case study examples.

4.1 Selection of the in-depth case studies

In an initial investigation possible nanotechnology application contexts were considered and qualitatively evaluated. Also studies to life cycle aspects of nanotechnology were analyzed. So far, only a handful of life cycle assessments (LCAs) on nanotechnologies have been completed. A summary of studies of life cycle aspects identified are provided.

Table 8. Overview of studies to life cycle aspects[48]

Nano-Product	Approach	Tech Benefits	Environmental Benefits	Reference
Nano-scale platinum-group metal (PGM) particles in automotive catalysts	Economic Input-Output Life Cycle Assessment (EIO-LCA)	reduced platinum-group metal (PGM) loading levels by 95%	overall reduced environmental impact	Lloyd et al. 2005
Clay-polypropylene nano-composite in light-duty vehicle body panels	Economic Input-Output Life Cycle Assessment (EIO-LCA)		overall reduced environmental impact; large energy savings	Lloyd and Lave 2003
carbon nanofiber	Ecobilan	reduced	NA (not com-	Volz and

---

[48] Scouce: based on Lekas (2005a) and own data

(CNF) reinforced polymers	TEAM software	weight, increased structural strength, improved conductivity	pared to traditional carbon fibers)	Olson 2004 (submitted)
Nano-scale platinum-group metal (PGM) particles in automotive catalysts	Eco-profile following LCA methodology	reduced platinum-group metal (PGM) loading levels by 50%	overall reduced environmental impact (10-40%)	Steinfeldt et al. 2003
Photovoltaic, dye photovoltaic cells as compared with multicrystalline silicon solar cells	Eco-profile following LCA methodology	Farbstoffzelle mit besserer Energy payback time aber geringerer Wirkungsgrad		Steinfeldt et al. 2003
Carbon nanotubes	substance flow analysis			Lekas 2005b
Energy consumption during nanoparticle production (TiO2, ZrO2)	Energy assessment			Osterwalder et al. 2006
Anti-reflex glass for solar applications as compared with traditional glass	Not Assessment, only indication of the environmental benefit	increased solar transmission	6% higher energy efficiency	BINE 2002
Printed circuit boards, Organic metal as compared with stannous/lead	Not Assessment, only indication of the environmental benefit	allows for application of a thinner coating layer with same functionality	10x more resource efficient	Omercon o.J.
Desanilation, Flow-through Capacitor as compared with resorved osmosis and destillation	Not Assessment, only indication of the environmental benefit	increased energy efficiency	very high energy efficiency	UBA 2006
Car tire, Si02, Carbon black	Not Assessment, only indication of the environmental benefit	increased road resistance	up to 10% lower fuel consumption	UBA 2006

Our goal in the selection process was to cover the spectrum of nanotechnological applications (a variety of manufacturing methods and basic nanoscale structures) as broadly as possible and address a diverse selection of research interests. With this in mind, selection of the case studies was made according to the following categories and associated criteria:
1. Type and scope of environmental impact
   anticipated eco-efficiency potential (high – low)
   potential for possible risks and/or toxicity (high – low)
2. Extent of market proximity
   State of development (already on market – still in long-term development)
   Market relevance (high – low)
   Potential application spectrum (wide – narrow)
3. Type of innovation
   Degree of innovation (small – large)
   Production turnover or volume (high – low)

In selecting the actual application contexts we chose to focus on specific issues/topics. Out of the entire spectrum of nanotechnological applications, four case studies with anticipatable eco-efficiency potential were specifically selected. Integrated technological problem solving innovations were our focus (cf. Kemp 1997, Huber 2004). The possible risks and potential dangers of nanotechnology applications, specifically the issue of nanoparticles, were analyzed and addressed. As a result of these deliberations, five application contexts with corresponding goals for the in-depth case studies were selected. The project results of these case studies follow.

Table 9. Overview of the case studies investigated

Application context	Goal
Eco-efficient nanocoatings	Presentation of the eco-efficiency potential of nanocoatings in the form of a comparative ecological profile (Nanocoating based on sol-gel technology as compared with waterborne, solventborne, and powder coat industrial coatings)
Nanotechnological process innovation in styrene production	Presentation of the eco-efficiency potential of nanotechnology in catalytic applications in the form of a comparative ecological profile (Nanotube catalyst as compared with iron oxide–based catalysts)
Nanotechnological innovation in the video display field	Assessment of eco-efficiency potential in video display development by means of a qualitative comparison (Organic LED displays and nanotube field emitter displays as compared with CRT, liquid-crystal, and plasma screens)
Nano-applications in the lighting industry	Presentation of eco-efficiency potential of nano-applications in the lighting industry in the form of a comparative ecological profile (White LED and quantum dots as compared with incandescent lamps and compact fluorescents)
Potential risks of nanotechnological applications	Discussion of possible risks and harzards using titanium dioxide as an example; less a consideration of environmental impact

4.2 Case study 1: Eco-efficient nanocoatings

4.2.1 Content, goal, and methods of the case study

In the course of this case study the ecological potential of nanotechnology-based coating processes is investigated. As a specific example, we look at a potential process for coating aluminum that utilizes a new nanocoating based on sol-gel technology. Evaluation of the ecological relevance is carried out by means of a comparative life cycle assessment.

The investigation of the ecological aspects are carried out by comparison with other industrial coating systems, specifically waterborne, solventborne, and powder coatings, including associated pre-treatments.

4.2.2 Scope of the investigation

4.2.2.1 Fundamentals of surface coating technologies and their relevance for the environment

The industrial application of self-curing or curable organic compounds to a surface is referred to as a surface coating process. These compounds are applied in very thin layers. By means of a chemical reaction or other physical process, a binding, durable film is created. As part of the surface coating process, one must also consider the pre-treatment of the surface as well as any follow-up treatment. The essential steps in the process are:
- Pretreatment (blasting, grinding/sanding, degreasing, deoxidizing/pickling, phosphating, chromodizing, passivation)
- Application of the surface coating
- Aftertreatment/curing

All liquid coatings share a similar basic make-up that consists of four primary components:
- Binder
- Liquid carrier or solvent
- Coloring agent (organic or inorganic pigments)
- Fillers and additives

Pretreatment
The surface of the component to be coated must be thoroughly cleaned of all grease and dust particles by means of physical or chemical processes. Improvements to corrosion resistance and the later surface coating bond are also part of the pretreatment process. Typical pretreatments include:
- for plain carbon steel: degreasing, pickling, phosphating, passivation
- for aluminum: degreasing, deoxidation, chromodizing or the application of a non-chromated conversion coating based on zirconium, molybdenum, titanium, or silicate

The surfaces of aluminum components generally receive a chromate conversion treatment. Following degreasing and an acid wash the components are placed in a chromate bath. The bath is strongly acidic (pH value 1 to 2) and has as its most important ingredients chromic acid and complex fluorides. As a protection against corrosion, zinc surfaces are also chromated in order to prevent the formation of white rust. In some cases, mechanical processing (for example the sanding of damaged paint areas on a motor vehicle) can also be part of the preparation process. Often an undercoat or primer is applied and serves as a bond between the component material and the surface coating. The undercoating also provides protection against corrosion and consists of chromates as well as zinc and lead compounds.

Chromating
Until recently, chromate treatment of aluminum was indispensable for the adhesion of a surface coating. With aluminum in particular, exposure to the oxygen in the air quickly causes a thin aluminum oxide coating to form. In order to make possible the adhesion of the surface coating to this oxide layer, the metal parts have always been treated with chromic acid. This reduces the thickness of the existing oxide layer, eliminates other metallic particles, and causes the formation of a new mixed aluminum–chrome oxide layer. The result is a corrosion-resistant aluminum surface and the formation of a protective coating in preparation for the following surface coat application. The chromate conversion process, because of the harmful environmental effect of its chromium compounds, is a significant problem.

The toxicology of the chromium compounds in use is also a problem: Chromic acid is extremely corrosive. The chromate treatment process produces waste water containing chromium(III) and chromium(VI), whose removal requires extensive waste water treatment processing. Substances containing chrome(VI) are considered particularly poisonous and are furthermore carcinogenic. Sludge containing chromium(VI) must be disposed of at expensive hazardous waste disposal sites (Anger 1982; Funk 2003). As a consequence of the EU end-of-life vehicle directive (2000/53/EG), beginning in 2007 the use of chromium-containing products in the pretreatment of parts to be surface coated will therefore no longer be permitted.

The search for alternative corrosion-resistant coatings for aluminum parts has therefore been underway for sometime. A number of processes, for example, pre-anodizing, are being explored; nonetheless, the corrosion protection these alternatives offer is, so far, not equal to that of chromating.

A coating based on organic-inorganic polymers developed by Nano Tech Coatings GmbH is of particular interest. The use of this surface coating fully obviates the need of chromating or similar pretreatments.

The surface coating process
Industrial surface coatings generally consist of a binder, pigment, solvent and additives. The composition of the binders (resins) determines the properties of the surface coating, including adhesion, mechanical and chemical resistance, sheen, and weather resistance. Pigments and additives are chiefly responsible for properties of color. They may also provide protection from corrosion (for example, zinc dust, red lead oxide) or include

stabilizers for UV protection. Pigments used in undercoatings and primers also serve to cover surface irregularities (filler). Additives are also used to improve application and performance properties. During the drying phase, the solvents become evaporative, volatile substances. Their purpose is to keep the solid components of the coating dissolved or dispersed and to maintain the applicable consistency and working properties of each coating.

Environmental impact of paint and surface coatings
A 1995 emissions study looked at the twelve business sectors most important for the application of industrial surface coatings – not including mass production surface coating operations in the automotive industry (Mink & Rzepka 1995). According to the study, 335,000 tons of solvent were being emitted annually. In comparison, the emissions from surface coating operations in automobile production are quite low, perhaps on the order of 30,000 tons annually, corresponding roughly to the emissions from automotive paint repair industry. The paint lines in the automobile industry already demonstrate a very high level of environmental awareness.

If automobile industry figures are included, annual emissions are close to 370,000 tons. This makes up roughly 40% of total solvent emissions or 15% of VOC emissions.

4.2.2.2 Specific aspects of the investigation and its scope

The study looked specifically at the surface treatment of light-alloy parts such as those increasingly being employed in automobile manufacturing. The surface treatment of a 1 sqm aluminum automobile part with various clear lacquers served as a functional reference unit.

The system boundaries included the entire life cycle of the coating product, including pretreatment of the surface (see Fig. 13). The individual life cycle stages in this case are :
- Procurement of raw materials
- Manufacture of primary materials (binder, solvent, etc.)
- Manufacture of surface coating
- Surface pretreatment
- Application (surface coating operation)
- Utilization phase
- Disposal/recycling

For the comparative profile, it should be noted that the surface pretreatment processing step could only be dealt with on a qualitative basis due to gaps in the data; the last phase, disposal/recycling, is in every case assumed to be identical and was not considered. The most relevant environmental impact criteria was expected in the surface coating production stages, including manufacture of raw materials, pretreatment, and application, as well as in the utilization stage.

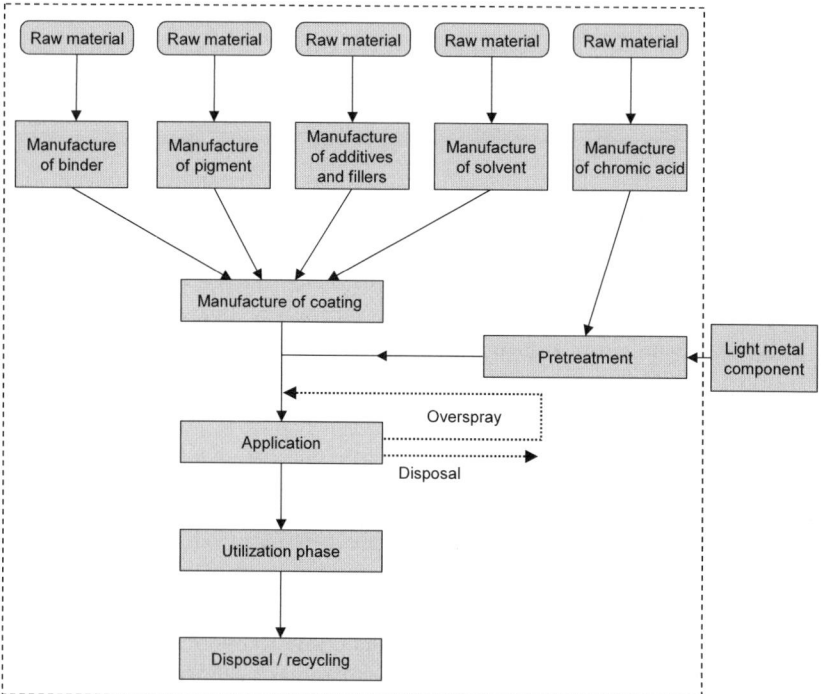

Fig. 13. System limits for the comparative life cycle assessment

4.2.2.3 Selection of variants

Criteria for the selection of the variants:
- Differentiation by deployment of the basic material forms (binder, etc.)
- Differentiation according to type of surface pretreatment required
- Differentiation according to method of application

For the environmental significance of the surface coating systems being considered, the following influential parameters were ascertained:
- Composition of surface coating
- Required surface pretreatment
- Coating thickness

Using these parameters it was possible to derive four variants that currently reflect the state of the technology. The so-called nanocoating is treated as the fifth variant; however, unlike the other four variants, this has not yet been implemented in automotive production surface coating applications.

Variant 1 and 2: single- and dual-component clearcoats (conventional clearcoat)
Among the clearcoat finishes under consideration, the single-component clearcoat (1K CC) has been in use the longest. It therefore brings with it a great deal of working knowledge and a greater degree of development. Dual-component clearcoat (2K CC) is principally used in applications demanding a greater degree of quality and durability. In the automotive branch, both coatings are increasingly being replaced by waterborne and powder coatings, as they are no longer able to do justice to the high standards of the Technical Guidelines on Air Quality Control. The solvent percentage in both coating systems is roughly 50%. Various (synthetic) resins are use as binders in both systems. The percentage of additives, however, is relatively small and consists mainly of flow control agents and light stabilizers.

Variant 3: Waterborne clearcoat
Waterborne clearcoat is the most frequently utilized surface coating process in the automotive industry. The amount of solvent used in waterborne clearcoats is higher than in conventional lacquers, but the primary solvent is water (39.8%), which fully evaporates during the drying phase and therefore harmless to the environment. The use of water rather than volatile solvents as the liquid component also makes a difference in primary energy demand: 62.9 MJ/kg lies far below that of the conventional and powder coatings.

Variant 4: Powder clearcoat
Unlike all other surface coatings, the powder coat process utilizes no solvents or other liquid carriers and is therefore considered to be particularly safe for the environment. Significant efforts are presently being made to utilize powder coating in more and more application areas. Lack of a liquid carrier also brings with it some disadvantages: For example, dip or immersion application is, of course, impossible.

Variant 5: Nanoparticle clearcoat from the firm NTC
The newly developed nanoparticle coating differs in many respects from conventional coatings. The process is fundamentally different. Like traditional liquid clearcoats it consists of a binder, a liquid carrier, fillers, and additives. The binder, however, does not have the usual organic structure, but is instead a so-called inorganic-organic hybrid polymer. The nanoparticle coating is manufactured by means of the sol-gel process. The sol-gel process has been in existence for a long time, however through increased research and development activities in recent years it has gained in importance. It is viewed as an especially promising field of nanotechnology.

The nanoparticle coating can be applied using customary methods. During the drying phase of the "sol" (the coating in its liquid phase), the particles suspended in it join together to form the so-called "gel." The material is heated to a temperature of 160°C, the liquid carrier fully evaporates from the layer and the particles bind together to form a stabile polymer network (Van Ooij et al. 2002). Significant advantages of this process include: a thinner coating providing the same functionality and the elimination of the chromate pretreatment, which is no longer necessary.

Table 10. Overview of the variants considered[49]

		Variant 1: 1K CC	Variant 2: 2K CC	Variant 3: Waterborne clearcoat	Variant 4: Powder coat	Variant 5: Nanocoat
Composition	Binder	46.7%	51.5%	41.6%	95%	55%
	Additives	2.6%	3.5%	1.4%	4.6%	3%
	Liquid carrier	50.7%	45%	57%, of which water makes up: 39.8%	0.4%	42%
Surface pretreatment	Processes utilized	Chromating phosphating anodizing	Chromating phosphating anodizing	Chromating phosphating anodizing	Chromating phosphating anodizing	Mild alkaline rinse
	Materials utilized (chromating)	Chrome (VI) or chrome (III)	Chrome (VI) or chrome (III)	Chrome (VI) or chrome (III)	Chrome (VI) or chrome (III)	Mild alkaline rinse
	Number of baths	10-12	10-12	10-12	10-12	3
	Coating thickness	35 µm	35 µm	35 µm	65 µm	5 µm

---

[49] Source: Harsch and Schuckert (1996) and authors

Primary energy requirement	87 MJ/kg	97.7 MJ/kg	62.9 MJ/kg	124 MJ/kg

4.2.2.4 Description of the life cycle stages and base data utilized

The study "Comprehensive Assessment of Powder Coat Technology as Compared to Other Surface Coating Technologies," by Harsch and Schuckert (1996), provided the base data for further investigation of the life-cycle stages; this study provides a life cycle assessment of the powder coating process as compared to other industrial surface coating technologies.

Production and manufacture of raw materials (surface coating components)

Chromic acid: Chrome ore is the raw material from which chrome(VI)-containing products for chromating are manufactured. South Africa, with between 30% (1992) and 42% (1996) of the market, is the largest worldwide producer of chrome ore, followed by Kazakhstan, Turkey, and India. In 1996, the six largest producers yielded roughly 86% of the approximately 12 million metric tons that were extracted worldwide.

Chrome ore is chiefly used in iron and steel metallurgy to produce stainless steel having special properties. In the chemical industry chrome ore is used to produce numerous compounds for application in diverse areas. Among these, chrome(VI) (chromium trioxide CrO_3), used for chromating. In 1993, according to the U.S. Bureau of Mines, roughly 77% of the chrome used in the OECD (Organization for Economic Cooperation and Development) countries was used in metallurgy, 9% in the fireproofing industry, and 14% in the chemical industry. The market segment belonging to the chemical industry – which includes the manufacture of chromic acid – is shrinking due to environmental protection concerns.

Binders
A large number of organic and synthetic resins are utilized in binders for industrial surface coatings. These binder compounds are utilized in various proportions and ratios, as needed, in all industrial surface coatings.

- Acrylic resins are synthesized from various primary products
- Epoxy resins are manufactured by means of a condensation process and serve to produce an especially durable and stable surface coating process
- Polyurethane resins are produced from polyether and polyester and have outstanding surface characteristics
- Polyester resins are condensation products resulting from saturated monomers
- Alkyd resins are produced from polyvalent alcohols and polycarboxylic acids
- Melamine resins are synthesized from the source materials melamine and formaldehyde

Hardeners
- Isophoron diisocyanate (IPDI)
- Hexamethylene diisocyanate (HDI)
- Phthalic anhydride (PA) with powder coating
- Triglycidylisocyanurate (TGIC)
- Silane in conjunction with nanocoat

An overview of the binders and hardeners in use is given in Table 11.

Table 11. Binders and hardeners utilized in the variants under consideration[50]

	Variant 1: 1K CC	Variant 2: 2K CC	Variant 3: Waterborne clearcoat	Variant 4: Powder coat	Variant 5: Nanocoat
Proportion of binder and hardener	46.7%	51.5%	41.6%	95%	55%
Binder utilized	Acrylic, melamin, and polyurethane resin	Acrylic resin	Acrylic, melamin, saturated polyester resin	Acrylic resin	Epoxy resin (proportion: 50%)
Hardener utilized		HDI prepolymer		Acid catalyst (PA)	Silane (proportion: 50%)

---

[50] Source: Harsch and Schuckert (1996) and authors

Solvent
Organic substances are the most common solvents. Before application they serve to maintain the coating in a liquid state. Following application the solvent evaporates and the coating becomes solid. Commonly used solvents include:
- Diacetone alcohol
- N-Methylpyrrollidone (NMP)
- Aromate
- Butyl acetate
- Butyl diglycol acetate
- N-butanol
- Secondary butyl alcohol (SBA)
- Butyl (poly)glycol

Pigments, fillers, and other additives are also used in very minimal quantities, and were not included in further calculations.

Manufacturing industrial surface coatings

Liquid coating systems
Manufacture of the coating is chiefly a process of mixing together the necessary components. Losses at this stage are minimal and can therefore be ignored. In a premixer the binder, liquid carrier or solvent, pigment, and fillers are mixed together and then ground. Any remaining additives are then added to the mixture in a let-down tank. After filtering, the mixture is packaged.

Powder coat systems
In the manufacture of powder coatings, all ingredients are first carefully weighed out and then fed into an extruder. The mixture is repeatedly powdered in several stages and finally conveyed to a packaging facility. Losses in powder coating manufacture are greater than with liquid coatings and amount to roughly 2–5%.

Nanocoat system
The manufacture of a nanocoating varies not so much in the processes, but rather in the binder that is utilized. As a rule, the binders used in industrial coatings have an organic structure. But inorganic binders are also utilized in certain applications. Their advantage is in their hardness and chemical durability. But because of their serious drawbacks, including difficult ap-

plication and brittleness after hardening, these coating materials do not have an extensive field of application.

In nanotechnology-based coatings, the so-called inorganic-organic hybrid polymers come into application. These new binders are a mixture of organic and inorganic binders and bring together numerous advantages from both types of binder. As the nanocoating cures, the inorganic particles begin to form a glasslike network with cross-linked organic elements.

The fundamental chemical reaction behind the manufacture of nanocoatings is based on the sol-gel process. This process frequently utilizes silicon-organic compounds, the so-called silanes. The synthesis of these binders is achieved by the hydrolysis of alkoxysilane. In this case study, the product Dynasylan® Glymo from Degussa was utilized. This part of the reaction leads to the formation of the inorganic part of the binder.

At the same time the formation of the organic part of the binder takes place. Organic side chains on the silane compounds undergo reaction to organic chains. As this inorganic-organic network forms, the coating hardens.

The finish condition of the coating, however, remains as it was before hardening. The binder is available as a low-viscosity colloidal suspension, whose particles have a diameter of 40–50 nm. The solvent, at this point, contains unreacted silane, silanol, and partially formed polysiloxane. This colloidal system is chemically in the so-called sol state; followed, after application and hardening of the coating, by the gel state. The entire process is therefore referred to as the sol-gel process (Wagner o.J.).

Base data and assumptions: Summary life cycle assessment data from the study by Harsch and Schuckert (1996) provided the base data for the four existing surface coating technologies. Since no quantified data exists for the primary materials in the nanocoating, specifically for the silane that is used, the data from the dual-component coating was also used for the nanocoating and multiplied by a "safety factor" of 1.5. The base data utilized for each 1 kg of applicable coating are provided in the appendix (Table 43).

Surface pretreatment

As a rule, before application of the surface coating, steel and aluminum components must receive a series of successive surface treatments to protect against corrosion. This is achieved through the application of a chemical conversion coating. Chromating is the usual corrosion prevention process, particularly for aluminum. The pretreatment occurs by means of a dip or spray process utilizing chrome(VI)-containing products. Aluminum can also be pretreated using one of the following processes: chemical oxidation, phosphating, or anodic oxidation.

Chromating
Chromating forms a conversion coating of complex chromates. The characteristic layer thickness of the resulting protective coating is no more than 0.01 to 10 micrometers. The chromating process itself is quite complex and involved. In automated applications, for example treating the body shell of an automobile, the part being treated often must pass through ten or more in part repetitive baths. At a minimum, it requires the following steps:
- Solvent cleaning: Grease, oil, and other contaminants are removed from the surface by means of solvents and other chemical processes.
- Intermediate rinse: The solvent and cleaning chemicals are rinsed away with cold or warm water. This step is omitted if only organic solvents were used for cleaning.
- Activation: Use of the activator (for example nitrous or sulphuric acid) yields a crystalline layer whose structure is significantly finer.
- Chromating
- Rinse: The excess chromate and phosphate solutions are rinsed away with deionized water. The water temperature during the chromating process must not exceed 50–60°C.
- Passivation: The surface coating is passivated either by spraying or dipping in very dilute chromic or chromic/phosphoric acid solutions.
- Drying: The wet treatment is directly followed by a warm-air dry off. To avoid powdery or poorly adherent layers the temperature should not exceed 60°C for yellow or clear chromating or 80–90°C for the green chromate coating (Gersing o.J.).

The conversion coating consists of oxide hydrates of chrome(VI), chrome(III), and aluminum. The surface coating is typically assigned to one of three classes (Jelinek 1997):
Class 1: coating weight 3.2 to 11 g/m^2;
suitable for greatest corrosion resistance even without any further coating
Class 2: coating weight 1.1 to 3.8 g/m^2;
corrosion prevention and pretreatment before application of finish surface coat
Class 3: coating weight < 1 g/m^2;
for decorative purposes only, minimal corrosion protection
The properties – for example layer thickness and corrosion resistance – of the conversion coating produced by the chromate treatment are depend-

ent upon the temperature of the bath, the treatment time, and the solution chemistry of the chromate bath. An overview of the various bath chemistries is given in Table 12.

Table 12. Typical composition of various chromate baths[51]

Composition	Quantity (g/l)	Ph value	Temperature (°C)	Treatment time (min.)
Chromic acid	3 – 7	1.2 – 1.8	30 – 35	2 - 5
Na (or K) dichromate	3 – 6			
Potassic fluoride	0.5 - 1			
Chromic acid	3.5 – 4	1.5	30	3
Sodium dichromate	3 – 3.5			
sodium fluoride	0.3			
Chromic acid	5	1.8	30 – 35	2 – 3
Sodium dichromate	7			
Sodium fluoride	0.6			
Ammonium dichromate	10 – 350	< 3	30	1 - 5
Hydrogen fluoride	0.25 – 11			

Phosphating

Phosphating is commonly used as a pretreatment for steel and iron. But it is an alternative to chromate for the pretreatment of aluminum as well. Treating aluminum with a dilute phosphoric acid solution produces a thin film of aluminum phosphate. Compared to chromating and anodizing, phosphating is certainly more expensive and more complicated, but it results in a more corrosion-resistant surface, which ensures a better bond for surface coatings. Phosphate coatings solutions for aluminum generally have a composition similar to those used for treating steel. The main components include metal hydrogen phosphates, oxidizing agents and complex fluorides (Brock et al. 1998).

Anodizing

The anodizing process is comparable in number of baths and complexity to chromating. But it offers significant advantages with respect to environmental impact (chrome-free, fluoride-free). The chrome(VI)- and fluoride-containing solutions used in chromating are replaced by conventional sulphuric acid. This leads to a much higher quality water and waste discharge. Furthermore, the resulting conversion coating is more corrosion-resistant than in the chromate process.

---

[51] Source: Jelinek (1997)

4.2 Case study 1: Eco-efficient nanocoatings

With nanocoatings no extensive pretreatment is required. Only the very first stage, a mild alkaline rinse, is necessary.

Table 13. Comparison of the processes by number of baths[52]

Baths, number	Chromating	Anodizing	Pretreatment for nanocoating applications
1	Mild alkaline degreasing	Alkaline degreasing	Mild alkaline rinse
2	Rinse	Rinse	Rinse
3	Alkaline etching	Alkaline etching	
4	Rinse	Rinse	
5	Rinse	Rinse	
6	Pickling/activation	Pickling/activation	
7	Rinse	Rinse	
8	Chromating	Anodizing	
9	Rinse	2x rinse	
10	Deionized water rinse	Sealing	

Table 14. Environmental advantages and disadvantages of the processes[53]

<table>
<tr><th colspan="2"></th><th colspan="2">Yellow chromating</th><th colspan="2">Anodization</th><th colspan="2">Pretreatment for nanocoatings</th></tr>
<tr><td rowspan="5">Raw material</td><td>Compound utilized</td><td colspan="2">Chromic acid (CrO$_3$)</td><td colspan="2">Sulphuric acid (H$_2$SO$_4$)</td><td colspan="2">Mild alkaline cleaner (pH value 8–10)</td></tr>
<tr><td>Water Hazard Classification (WGK)</td><td colspan="2">WGK 3</td><td colspan="2">WGK 1</td><td colspan="2">WGK 2</td></tr>
<tr><td>Storage</td><td colspan="2">Hazardous substance storage, extensive legal and structural requirements</td><td colspan="2">Less complicated hazardous substance storage</td><td colspan="2">Simple storage</td></tr>
<tr><td>Handling instructions and safety data</td><td colspan="2">- Special instructions for the handling of chromic acid necessary (industrial safety)
- Extremely toxic to water organisms
- Carcinogenic</td><td colspan="2">- Special instructions for the handling of sulfuric acid necessary (industrial safety)
- Causes severe burns</td><td colspan="2">Health hazard only in the case of inhalation or ingestion</td></tr>
<tr><td>Application technique</td><td>Dipping</td><td>Spraying</td><td>Dipping</td><td>Spraying</td><td>Dipping</td><td>Spraying</td></tr>
<tr><td>Input / Output</td><td>Effluent volume</td><td>Significantly greater for dip bath technology</td><td>Can be minimized by selective spraying</td><td>Significantly greater for dip bath technology</td><td>Can be minimized by selective spraying</td><td>Significantly greater for dip bath technology</td><td>Can be minimized by selective spraying</td></tr>
</table>

[52] Source: Funk (2003) and authors
[53] Source: Funk (2003) and authors

Tech. / economic characteristics	Waste purity	$Cr^{3+/6+}$ very toxic		Acidic sulfate-containing solution		90% biodegradable	
	Exhaust air	Contains chromic acid vapor		Contains sulfuric acid vapor, which is fed back after condensation		No impact	
	Energy	Bath temperature 25–40°C, larger volume must be maintained at temp.	Bath temperature 25–40°C	Room temp. (RT) bath, cooling necessary at higher capacities	RT bath, cooling possibly not required	Only 3 baths necessary instead of the usual 10 to 12	
	Handling/cycle time	ca. 5 minutes		ca. 5 minutes			
	Floor space required	ca. 100 m²	ca. 20 m²	ca. 100 m²	ca. 20 m²	significantly less	
	Throughput	100–200 m²/h	200–400 m²/h	100–200 m²/h	200–400 m²/h		
	Coating thickness	3–8 µm		10 µm		none	
	Plant space requirements	ca. 20 m³ per bath and rinse	ca. 5–10 m³ per activation bath and ca. 5 m³ per rinse	ca. 20 m³ per bath and rinse	ca. 5–10 m³ per activation bath and ca. 5 m³ per rinse		
	Legal guidelines	2000/53/EG					

Base data and assumptions: Because appropriate, quantifiable data for the given pretreatment methods is not available, it is not possible to include this processing step in further environmental impact calculations. However, the comparison already makes clear that nanocoatings offer unique environmental advantages.

The surface-coating process (application)

Following pretreatment, the nano-clearcoat – particularly emphasized in this study – is applied in the same manner as a single-component clearcoat. The main difference to the other coating systems is that only a tenth of the usual coating material is needed. Because the application process for nanocoatings is the same, we will rely heavily on Harsch and Schuckert (1996).

The comparison of the various coating systems by application process is based on design data for new facilities of the same production capacity. The general steps of application are shown in Table 15. From this it can be seen that application generally consists of three steps: Interior, exterior shell, and manual touch-up.

Table 15. Technical stages of clearcoat application[54]

1K CC, 2 K CC, Nano-coating (NC)	Waterborne clearcoat	Powder clearcoat
Intermediate dryer	Intermediate dryer	Intermediate dryer
Cooling zone	Cooling zone	Cooling zone
Air lock	Air lock	Air lock
Coating	Coating	Coating
Inner body shell	Inner body shell	Inner body shell
Robot station	Robot station	Robot station
Coating	Coating	Coating
Outer body shell	Outer body shell	Outer body shell
ESTA station	ESTA station	Tribo station
Manual touch-up	Manual touch-up	Manual touch-up
Air lock	Air lock	Air lock
	Infrared predryer	
Dryer	Dryer	Dryer

The application process is influenced by certain variables that have an influence on quality and quantity of the output. These include formulation of coating, the coating system, energy source, and production capacity. For the purposes of this study average values for these variables were used. The application base data is available in the appendix (Table 44).

Table 16. Modifiable variables and their average characteristics for the various coating systems[55]

Coating system		1K CC	2K CC	Waterborne clearcoat	Powder coat	Nanocoat
Production capacity		\multicolumn{5}{c}{40 body shells per hour}				
Surface area coated		\multicolumn{5}{c}{20 m²}				
Coating application	Coating thickness	35 µm	35 µm	35 µm	65 µm	5 µm
	Application efficiency	63.6%	63.6%	61.5%	64.7%	63.6%
	Overspray recycling	no	no	no	no	no
	Qty. (kg/ body shell)		2.303	3.174	2.421	0.32

---

[54] Source: Harsch and Schuckert (1996) and authors
[55] Source: Harsch and Schuckert (1996) and authors

Use phase, disposal, and recycling

In the use phase the various environmental impacts of the variants studied – resulting from the various coating quantities applied – are determined. A useful life of 200,000 km was assumed for the automobile. The consumption reduction rule used by Harsch and Schuckert (1996) was likewise utilized.

The so-called consumption reduction rule says that given the weight and consumption of a specific automobile type, a 10% reduction of weight will result in a 2.5–6% reduction in consumption. In the area between 90 and 100% of the original total weight, it is assumed that weight saved and fuel saved are proportional. For the baseline consumption, an automobile type of average fuel consumption was selected from the GEMIS 4.1 database.

It is assumed that the disposal/end-of-life phase will not significantly vary and it is therefore not included in the assessment.

4.2.3 Life cycle inventory analysis

In the life cycle inventory analysis, the material and energy relationships between the coating systems being evaluated and the environment are noted, i.e. inputs from the environment and outputs into the environment are recorded. The goal of the life cycle inventory analysis is to establish a data inventory based on functional equivalents for the selected variants.

Since quantitative data for the pretreatment is not available and the disposal/recycling life-cycle stages were not considered, estimates were necessary with the various variants for the coating process, including raw material production, application, and the use phase. The following tables depict the calculated inputs and outputs for the variants studied.

Table 17. Quantities utilized for chromating and surface coating operations (g/m² of coated aluminum automobile surface area)[56]

	1 K CC	2 K CC	Waterborne clearcoat	Powder coat	Nanocoat
Chromating	2.50	2.50	2.50	2.50	0.00
Binders / hardeners	59.99	59.30	66.02	115.00	8.80
Additives	3.34	4.03	2.22	5.57	0.48
Solvent	65.12	51.82	27.30	0.48	6.72

[56] Source: Harsch and Schuckert (1996) and authors' calculations

Water	0.00	0.00	63.16	0.00	0.00
Taotal	130.95	117.65	161.20	123.55	16.00

Table 18. Total primary energy consumption (MJ/m² coated aluminum automobile surface area)[57]

	1 K CC	2 K CC	Waterborne clearcoat	Powder coat	Nanocoat
Coating production	11.175	11.250	9.982	15.010	2.345
Application	38.225	38.305	38.605	31.970	33.600
Use stage	7.37	7.37	7.66	13.87	1.02
Total	56.769	56.925	56.248	60.849	36.961

Table 19. Outputs relative to 1 m² coated aluminum automobile surface area[58]

		1 K CC	2 K CC	Waterborne clearcoat	Powder coat	Nanocoat
Air emissions						
NM VOC	[g]	4.835	4.722	4.323	5.780	1.590
Methane	[g]	5.821	5.796	5.899	6.109	3.989
NOx	[g]	5.894	5.793	5.029	5.968	3.279
SO_2	[g]	2.421	2.651	2.547	2.938	1.489
CO	[g]	10.472	10.486	10.778	18.944	1.869
Particulate	[mg]	626.099	743.811	598.365	548.658	484.093
HCl	[mg]	79.733	79.229	86.987	77.724	65.225
HF	[mg]	27.692	27.356	29.831	24.910	23.314
N_2O	[mg]	63.868	63.875	66.396	120.205	8.812
NH_3	[mg]	40.096	40.100	41.683	75.464	5.532
Water emissions						
Discharge water	[l]	198.255	195.246	203.430	171.742	165.012
CSB	[g]	1.054	1.405	1.242	1.046	0.419
TOC	[g]	0.956	1.123	1.005	0.972	0.519
BSB	[g]	0.240	0.171	0.232	0.270	0.054
Solids	[g]	0.491	1.142	0.508	0.537	0.492
HC	[mg]	14.409	12.612	10.969	17.101	3.374
NaCl	[g]	36.622	10.746	11.486	20.955	2.353

[57] Source: Harsch and Schuckert (1996) and authors' calculations
[58] Source: Harsch and Schuckert (1996) and authors' calculations

86   4 Assessment of sustainability effects in the context of specific applications

Iron [mg]	5.581	5.374	5.506	32.584	3.765
Nickel [mg]	0.225	0.200	0.260	0.454	0.086
Chromium [mg]	0.074	0.063	0.063	0.067	0.038
Lead [mg]	0.330	0.376	0.350	0.299	0.282
Copper [mg]	0.044	0.035	0.038	0.044	0.021
Cadmium [mg]	0.008	0.007	0.008	0.010	0.004
Waste					
Industrial waste [g]	7.964	7.087	8.227	21.806	2.928
Household waste [g]	2.274	1.747	2.275	2.158	0.486
Hazardous waste [g]	4.379	3.506	8.822	10.674	0.773
Radioactive waste [g]	0.227	0.230	0.221	0.345	0.104

The coating quantities are determined by the surface thickness to be applied. The minimal surface thickness of the nanocoating is the primary reason for the high resource efficiency. In spite of the relatively large amount of solvent, the absolute amount in the nanocoating is minimal as compared to the other variants.

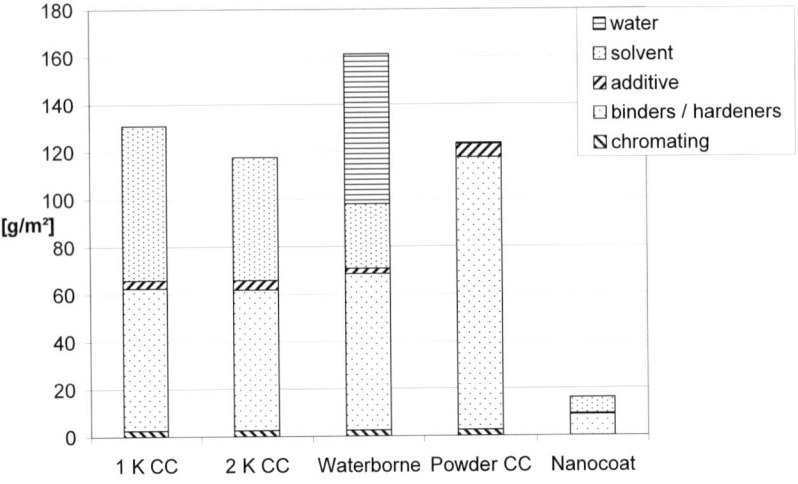

Fig. 14. Coating and chromating quantities (g/m² coated aluminum automobile surface area)[59]

---

[59] Source: authors (base data: authors, Harsch and Schuckert 1996)

The total primary energy consumption is primarily determined by the energy requirement of the application and includes not only the actual coating process but also energy expenditures for drying, etc. The differences here in the variants studied are minimal. The roughly 35% less total primary energy consumption of the nanocoating results is due to reduced quantity of coating material. However, during the use phase, the reduced mass also leads to savings in fuel consumption.

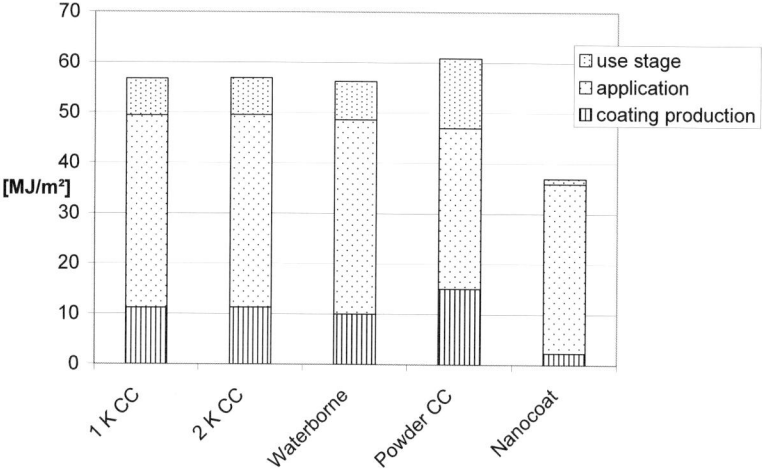

Fig. 15. Total primary energy consumption (MJ/m² coated aluminum automobile surface area)[60]

With the VOC emissions the advantages of the nanocoating are also quite evident, particularly in the manufacture and use life cycle phases. The VOC emissions of the nanocoating are ca. 65% lower than those of the other variants.

---

[60] Source: authors (base data: authors, Harsch and Schuckert 1996)

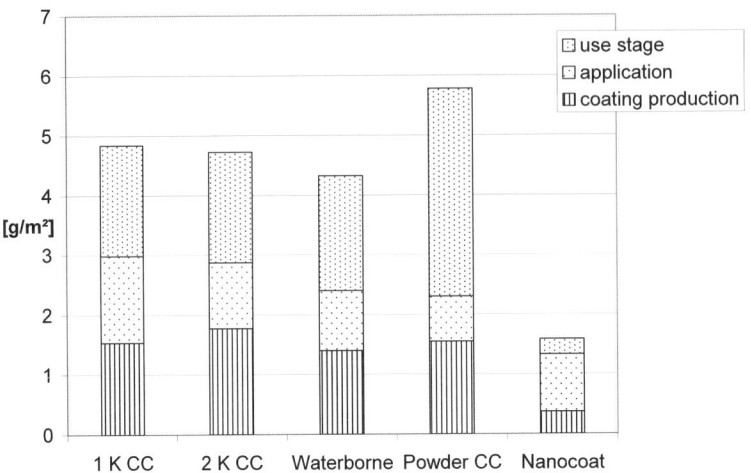

Fig. 16. VOC emissions (g/m² coated aluminum automobile surface area)[61]

The results for the other emissions values in Table 19 are similar. The reduction in waste generation also reflects the environmental advantages of the nanocoating.

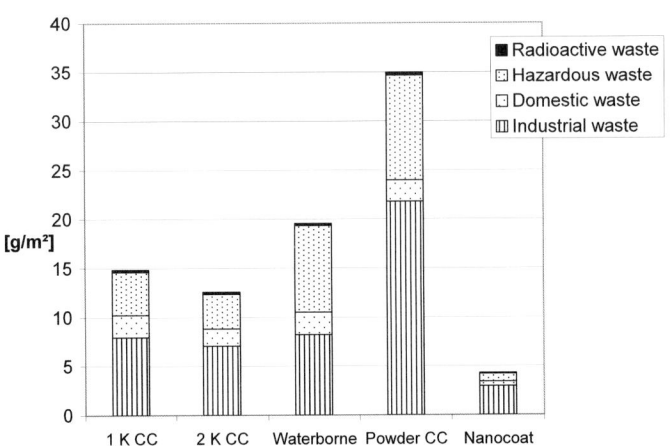

Fig. 17. Waste generation (g/m² coated aluminum automobile surface area)[62]

---

[61] Source: authors (base data: authors, Harsch and Schuckert 1996)
[62] Source: authors (base data: authors, Harsch and Schuckert 1996)

4.2.4 Life cycle impact assessment

To complete a life cycle impact assessment, it is necessary to have access to emissions data to which specific environmental impacts can be allocated. With the data available, it is only constructive to present an analysis for the impact categories greenhouse effect, acidification, and eutrophication.

4.2.4.1 Description of the environmental impacts

Greenhouse effect
The energy in the sunlight that strikes the surface of the earth in the course of the day is stored as thermal energy and then released at night as infrared radiation. A part of this infrared radiation is absorbed by trace gases in the troposphere (0–10 km) and reflected back to earth. This natural greenhouse effect is essential, otherwise the earth's surface would cool to inhospitable below-zero temperatures. The greenhouse effect we speak of as an environmental problem refers to the additional warming of the surface of the earth that is due to the increase in trace gases and the appearance of new greenhouse gases in the troposphere, for example HFCs (fluorocarbons). The most important greenhouse gases are carbon dioxide, methane, ozone, HFCs, and nitrous oxide, which arise to 50% from energy consumption, 20% from the chemistry industry, 15% from agriculture, and another 15% from the destruction of rain forests.

The greenhouse effect covers a wide range of effects that result from the warming of the earth's atmosphere. Among these are not only the rising mean sea level, but also the increase in extreme climatic weather conditions such as hurricanes, storm floods, catastrophic drought, etc. Changes in the composition and the range of flora and fauna are also already being looked at.

Acidification
Acidification is a collective term referring to several effects. The phenomenon can be primarily traced back to sulfur dioxide and nitrogen oxide emissions from the burning of fossil fuels in power plants and increasingly in motorized transport. In addition, ammonia, hydrogen chloride, and hydrogen fluoride emissions also contribute to acid rain. Sulfur dioxide and nitrogen oxide emissions react with atmospheric oxygen and water to produce sulfuric and nitric acid.

In the life cycle impact assessment, acidification potentials (AP) were generated using coefficients from a study by the Center for Environmental Science (CML) in the Netherlands (Heijungs 1992). Substances that only contribute to acid rain after oxidation (e.g. ammonia) or hydrolysis (e.g. SO2) are likewise included. In the CML models, only air emissions are considered; water emissions do not enter into the calculations.

Eutrophication
Eutrophication describes the spread of chemical nutrients into water bodies. The anthropological contributions of nitrogen compounds (e.g. nitrates) from excessive applications of fertilizers as well as phosphorus compounds (e.g. phosphates) from detergents or agricultural runoff lead to overfertilization of waters. In addition to these two groups of compounds, the COD (chemical oxygen demand) is enlisted as a measure for calculating organic pollutants. A consequence of the excessive nutrient enrichment is the appearance of vast algae growths. Dying algae decompose under a high degree of oxygen consumption and therefore lead to a shortage of oxygen in the water body. Decomposition and decay processes are the result and produce toxic substance such as hydrogen sulfide, which in turn leads to fish die-off. This means that the increased nutrients in our waters will do long-lasting and in part irreparable damage to a fragile ecological structure.

For the life cycle impact assessment the following factors were explored.

Table 20. Impact factors utilized

Air emissions	Greenhouse potential (GWP 100)	Acidification potentials (AP)	Eutrophication potentials (NP)
Carbon dioxide (CO_2)	1		
Methane (CH_4)	21		
Nitric oxide (NO_x)		0.70	0.13
Nitrous oxide (N_2O)	310		
Sulfur dioxide (SO_2)		1.00	
Hydrochloric acid (HCl)		0.88	
Hydrogen fluoride (HF)		1.60	
Ammonia (NH_3)		1.88	

4.2.4.2 Quantitative analysis of the life cycle impact assessment

The greenhouse effect is addressed in the investigation through emissions of carbon dioxide (CO2), methane, (CH4), and nitrous oxide (N2O); in this, the carbon dioxide emissions from the use of fossil fuel energy sources make up the greatest portion. Much as with total primary energy consumption, the nanocoating comes out roughly 1/3 better than the other coating variants.

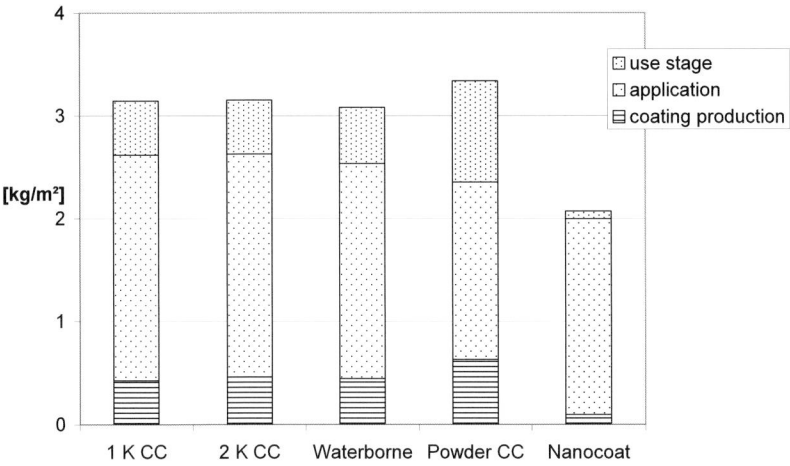

Fig. 18. Greenhouse potential (kg/m² coated aluminum automobile surface area)[63]

The acidification is addressed through emissions of nitrogen oxides, sulfur dioxide, ammonia, hydrochloric acid, and hydrogen fluoride, whereby the last three substances play a lesser role in the scenarios being looked at.

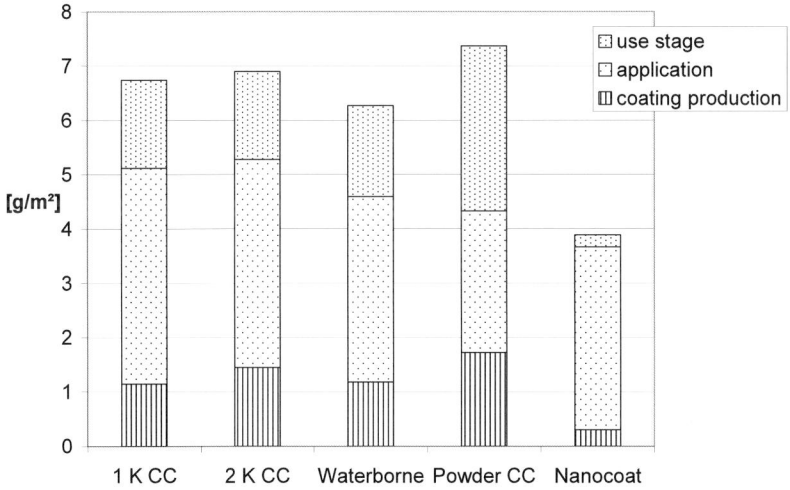

Fig. 19. Acidification potential (g/m² coated aluminum automobile surface area)[64]

---

[63] Source: authors (base data: authors, Harsch and Schuckert 1996)

In addition to the phosphorus compounds in detergents, nitrogen oxide emissions and organic pollutants, which are released due to the chemical oxygen demand (COD), also contribute to eutrophication. In our case nitrogen oxide emissions can primarily be held responsible. In this impact category as well, the nanocoating makes clear its advantage, ranking 40% better than the other variants.

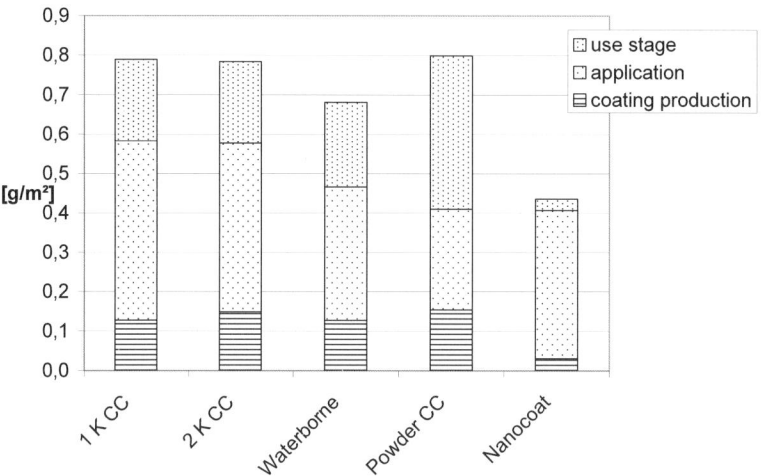

Fig. 20. Eutrophication potential (g/m² coated aluminum automobile surface area)[65]

4.2.4.3 Qualitative aspects of the impact assessment

A generally accepted quantitative process for representing ecological and human toxicity in a life cycle analysis does not exist. Therefore at this point we take a brief qualitative look at individual substances that demonstrate an impact on ecological and human toxicity. Because of the methods, no local or time-independent evidence could be given in the impact assessment.

Those substances that are of global significance beyond their source of emission should be recorded. Furthermore, substances without an effectivity threshold should be included. The goal should be to minimize or replace such substances to the extent possible.

---

[64] Source: authors (base data: authors, Harsch and Schuckert 1996)
[65] Source: authors (base data: authors, Harsch and Schuckert 1996)

Carbon monoxide (classified as mutagenic) has long been a problem for atmospheric pollution, particularly emissions in the transport sector. The successful reductions in CO achieved through the use of catalysts has made possible a tremendous reduction in emissions, such that the Federal Environment Agency, in a publication on technological options for the reduction of the impact of transport, came to the result that carbon monoxide no longer represents an air quality problem (UBA1999).

4.2.5 Case-study Summary

This case study makes impressively clear that in the area of surface coatings with respect to all emissions and environmental considerations studied that the utilization of nanotechnologically based coatings offers very great eco-efficiency potentials. Beyond this, the further advantages of simplified pretreatment were at least qualitatively shown. The minimal thickness necessary for such coatings leads to greater efficiency in use of resources; advantages in the usage phase can particularly be expected in the transport sector in the course of the trend to lightweight construction. In addition to the automotive industry, this potential would have an even greater effect on the airline and rail industries. A further potential for optimization can be found in the reduction of the solvent quota in nanocoating applications.

4.3 Case study 2: Nanotechnology Innovation in Styrene Production

4.3.1 Contents, goals and methods of the case study

Catalytic processes are among the earliest applications of nanotechnology. In principle, continuing gains in the efficiency and cost-effectiveness of catalytic processes can be achieved through the use of ever-smaller nanoparticles. Our understanding of catalysis is radically changing: the development of catalysts once relied upon empirical methods and values based on experience, but today the application of nanoanalytic methods, such as scanning tunneling microscopy (STM), makes it possible to reveal in detail the mechanisms of catalytic reaction.

In the course of this case study, we will investigate and look at an example of the ecological impact of nanotechnology-based catalytic applications. As a specific example, we will look at the application of a nanostructured catalyst utilizing nanotubes to the production of styrene. The ecological impact will be evaluated by means of a comparative life cycle assessment (LCA) of the specific processing stages as well as the entire styrene product life cycle. Investigation of the ecological aspects will be made by means of comparisons to existing chemical processes for styrene synthesis.

4.3.2 Scope of the investigation

4.3.2.1 Overview: Catalytic processes and nanotechnology

Catalysts are involved in the production of a great number of articles in everyday use. Catalysts are utilized in refining oil, setting free the energy in batteries and fuel cells, producing medicine and agrochemicals, plastics and paints, and in quite a number of environmental applications, for example the three-way catalytic converter in the automobile.

The generally accepted definition of a catalyst comes from Wilhelm Ostwald, who was awarded the Nobel prize in 1909 for his work on catalysis: "A catalyst is an agent which increases the rate of a chemical reaction without being consumed by it and without altering the final state of the thermodynamic balance of this reaction."

This definition holds true for all catalysts; they differ in functionality, but not in their effect. There are three different types of catalysts, each with its own distinguishing characteristics:
1. Heterogeneous catalysts
2. Homogeneous catalysts
3. Biocatalysts

Heterogeneous catalysts are those that exist in a different phase than the reactants. In homogeneous catalysis, the catalyst is in the same phase as the reactants. This has the distinct advantage that a greater number of active catalyst molecules are therefore available, whereas in the case of a heterogeneous catalyst, only the surface molecules are active. Biocatalysts, also called enzymes, are the most wide-spread catalysts. Without them life would not be possible, as almost all processes in nature are controlled by biocatalysts.

Catalysis has far-reaching importance for the chemical industries: 90% of all chemical production processes are based on catalysis. More than 80% of the output of the chemical industry is achieved by means of processes that occur in the presence of catalysts. With a world-wide catalyst

4.3 Case study 2: Nanotechnology Innovation in Styrene Production 95

market presently estimated at about 12 million dollars (US), the value of the resulting products ranges, according to various estimates, from 1.2 to 6 trillion dollars. The German share of catalyst production on the world market, about 4%, is not commensurate, however, with the significant role the chemical industry plays in the German economy (Herrmann 2000).

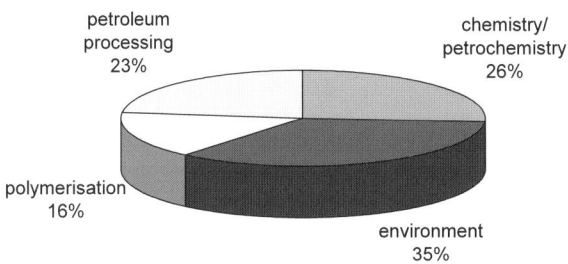

Fig. 21. Areas of application for catalysts[66]

Catalysts are among the oldest applications of nanotechnology. The catalytic converter, for example, has been a nanotechnology application from its very beginning. Nanotechnological progress in the development of new catalysts is already being achieved through improvements in the production of nano-scale particles. The reason for this is that the catalytic process is enhanced by these nanoparticles, which have a smaller diameter: catalytic reactions take place on catalytically reactive surfaces and the smaller the diameter, the greater the specific surface area in relation to volume. This yields significant advantages: the size of the catalyst can be reduced while retaining the same reactive surface area; likewise the number of reactive atoms on the surface can be increased for the same given mass. Along with progress in the production of nanoparticles, developments in the field of nanopermeable materials and nanostructured surfaces are also beneficial for catalysis.

This was shown in a study on technological development in catalytic converters for automobiles (Steinfeldt et al. 2003). Increasingly smaller and more homogeneous precious metal particles are being used in catalytic converters for automobiles; the improved catalytic action of these platinum

---

[66] Source: Herrmann (2000)

metal group particles (PGM) relies upon this surface-to-volume-ratio effect. This, along with continual improvements in the thermally stable bonding of these particles in the converter and reductions in the aging process, reduces the amount of PGM required and likewise makes it possible to meet ever stricter international emissions standards. Environmental relief is thus obtained through a reduction of pollutants in automotive exhaust emissions and through the reduced environmental impact achieved by "saving" PGM, the production of which is very expensive. Depending on the type of emission being considered, the eco-efficiency potential ranges from 10 – 40% (Steinfeldt et al. 2003).

Our understanding of catalysis is also changing radically. The development of catalysts once relied solely on empirical methods and experiential values. Nanoanalytical methods such as scanning tunneling microscopy enable us to investigate the mechanisms of catalytic reactions in ever-greater detail and to better understand and model them.

4.3.2.2 Focus of the investigation: styrene synthesis

Styrene production is considered to be one of the ten most important petrochemical processes and styrene is one of the most important base chemicals in the chemical industry. Numerous important synthetic polymer materials – the chief one being polystyrene – are made from the monomer styrene. Alongside ethylene and vinylchloride, it is one of the most important monomers.

Styrene: Economic impact and market data
Worldwide demand for styrene is estimated to be more than 20 million tons and growing by 5% annually (Rohden 2001). In 20 the styrene industry did face a 1.8% lapse in demand, but by the following year demand had again increased by 5.1% (Childre 2003). An overview of the development of styrene production is provided in Table 21 and Table 22.

4.3 Case study 2: Nanotechnology Innovation in Styrene Production

Table 21. Production data and capacity in Western Europe[67]

Styrene Production and Capacity in Western Europe (1000 tons)					
Year	2001	2002	2003	2004	2005
Capacity	5,552	5,357	5,742	6,137	6,147
Production	4,908	4,965	5,190	5,500	5,510

Table 22. Styrene production worldwide[68]

Styrene world production (1000 tons)					
Year	1997	1998	1999	2000	2001
Production	17,600	18,400	19,100	19,900	20,400

Styrene is solely used to produce polymeric products, above all polystyrene. Other possible applications include the styrolacrylnitril copolymer (SAN), terpolymer from acrylnitril, butadiene and styrene (ABS), SBR composition rubbers and unsaturated polyester resins. However, the proportion of polystyrene being used in manufacturing is decreasing. In 1975, 60% the product groups worldwide belonged to polystyrene, but today it is only about 50% (ISEF 2001). As with all other petrochemical derivatives, styrene is very much dependent upon crude oil prices and thus is subject to major price fluctuations.

Description of styrene production
Styrene is produced in a multi-stage process that can be traced back all the way to the procurement of the raw material. The production of styrene is based on intermediate products obtained from petrochemical resources, and the entire process therefore stretches from the exploration, extracation, and refining of crude oil and gas to the synthesis of styrene in the refinery. Fig. 22 provides an overview of the entire production process.

Naphtha and other petroleum products are produced by atmospheric distillation under normal pressure and temperatures ranging from 350 to 370° C. Subsequently, long-chain saturated hydrocarbons of the naphtha are split by means of steam cracking into low-molecular compounds such as butane (C_4H_{10}), benzene (in small amounts), and ethylene (C_2H_4) . Ethylene, propene (propylene), butane and other products are likewise obtained from natural gas by cracking. Ethylbenzene is produced catalytically by alkylation of benzene with ethene (using $AlCl_3$ or silica gel) under pres-

---
[67] Source: CEFIC (2006)
[68] Source: VCI (2003)

sure. It is an aqueous phase ethylation at an ethylene-benzene ratio of about 0.6/1, benzene conversion of 52–55%, temperature of 85–95°C, and atmospheric or slightly greater pressure. Styrene ($C_6H_5C_2H_3$) is produced from ethylbenzene ($C_6H_5C_2H_5$) by means of a catalyst. In a further processing step, the styrene is converted by polymerization to polystyrene, thus forming the base material for the production of numerous synthetic materials.

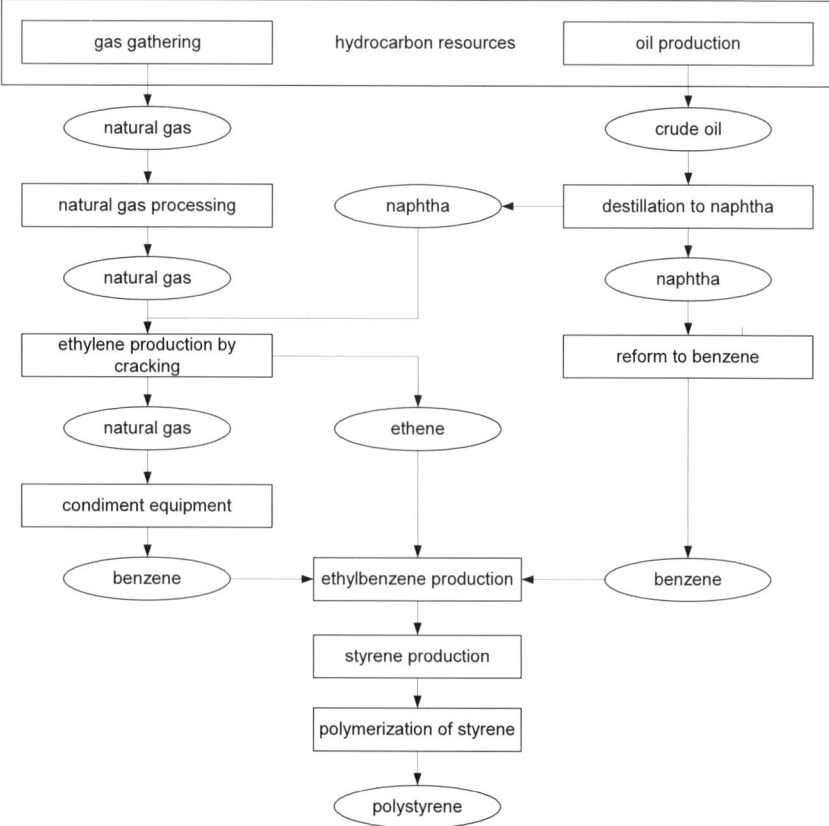

Fig. 22. Stages of industrial production of styrene and further processing for polystyrene[69]

---

[69] Source: APME (1997)

4.3.3 Scope of investigation and accessibility of data

The scope of this case study is not specifically the entire styrene product life cycle, but rather the stages of processing in the production of styrene. However, it does make reference to the entire styrene product life cycle, i.e. the stages of raw material extraction, pre-production, and production of styrene are looked at in order to be able to assess the eco-efficiency potential within the overall plan.
The functional reference quantity for the comparison is 1 kg of styrene.
Comprehensive topical data on material and energy flow in the production of styrene are available from the Association of Plastics Manufacturers in Europe (APME). This includes summary life cycle assessment data for the classic production of styrene, i.e. data on all stages, from crude oil and natural gas extraction and processing to the actual production of styrene (APME 1999). Moreover, APME also provides life cycle assessment data on the intermediate products benzene and ethylene.
Between these intermediate products and the product styrene are two more stages: the ethylbenzene process and the styrene process, with the consequence that no clear differentiation is possible by means of this database.
There are further life cycle assessment data available in the Gabi materials database (Gabi 4 Datenbank 1999b) likewise for the overall styrene production process, but also for the ethylbenzene process (from a production plant in the Netherlands, Gabi 4 Datenbank 1999a). Unfortunately, comparison of all available data sets did not reveal sufficient congruence for the emission data specific to the styrene process. This may be due to differences in calculation procedures or different data sources. Therefore only a differentiated evaluation of the energy consumption in styrene production was possible, as well as specific estimates of individual material flows (heavy metals).
No quantified process data is available for the alternative styrene process based on a carbon-nanotube catalyst. However, assumptions about energy consumption can be derived from the description of the technology.

4.3.3.1 Selection of the variants

The production of styrene is a chemical process whose efficiency very much depends on the utilization of suitable catalysts. In contrast to the technologically established use of iron oxide−based catalysts, a newly developed catalyst based on nanostructured carbon tubes, so-called nanotubes, is now available.

Variant 1: Classical styrene synthesis using an iron oxide catalyst
Styrene is synthesized in an industrial process that has been known for roughly 60 years: the dehydrogenation of ethylbenzene. The dehydrogenation of ethylbenzene to produce styrene is a reversible endothermic balance reaction (Lieb & Hildebrand 1982):

$C_6H_5C_2H_5 \leftrightarrow C_6H_5C_2H_3 + H_2 \quad \Delta H^{600°C} = 124.9$ kJ/mol (1)

The dehydrogenation of ethylbenzene to produce styrene takes place at temperatures of 600°C in an ethylbenzene-water mixture and is assisted by potassium-promoted iron oxide catalysts. Along with this main reaction, other reactions take place; these are listed in the appendix (Fig. 53).

Two technologically different processes are used in the industry; they differ primarily in the way heat for the endothermic reaction is applied (Lieb & Hildebrand 1982; Schoen 2002). In more than 75% of the styrene production plants operating worldwide, dehydrogenation is carried out adiabatically. Multi-stage reactors or a reactor beds in series are used.

a. The adiabatic process developed by Dow Chemical Company:

The necessary heat for the reaction is applied in the form of superheated steam (approx. 830°C), which is added to the ethylbenzene steam before the catalyst. The mass ratio of ethylbenzene to superheated steam is between 1.5:1 and 2:1. The temperature of the reactive mixture as it enters the reactor is approx. 650°C and 570°C as it leaves. The styrene output after this first stage is still relatively low; therefore the reactive mixture is reheated to 640°C in a second (and if required in a third) stage and is dehydrogenated once more. The production capacity of such a plant is presently about 500,000 t/year.

4.3 Case study 2: Nanotechnology Innovation in Styrene Production

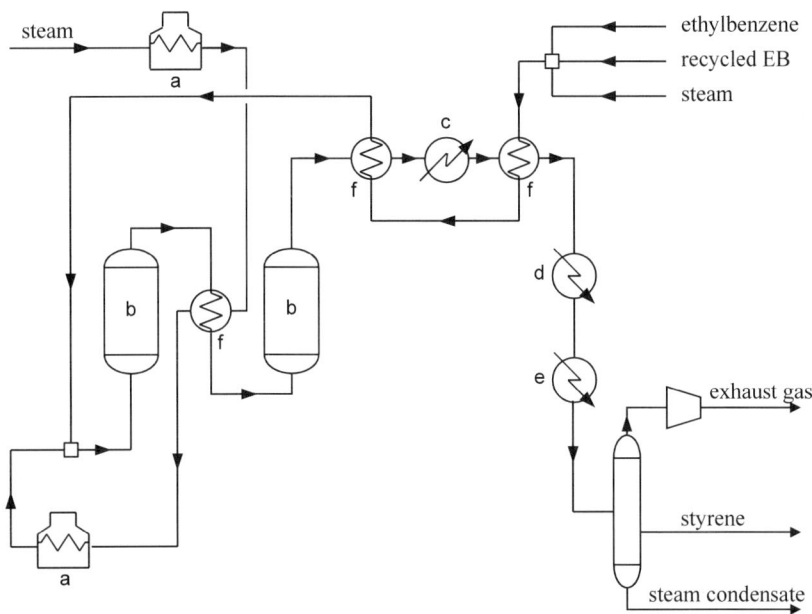

Fig. 23. Process schematic for the adiabatic dehydrogenation of ethylbenzene (EB)[70]
a) steam superheater; b) reactor; c) high-pressure steam; d) low-pressure steam; e) condenser; f) heat exchanger

b. The isothermal dehydrogenation process developed by BASF:

In this procedure, a constant temperature is maintained in tubular reactors that are heated by circulating flue gas or molten salt. The feed temperature of the reactive mixture is at about 600°C and can be kept nearly constant during the reaction with the catalyst layer. In this way the amount of superheated steam needed can be kept to about half of that used in the adiabatic process. Newly built dehydrogenation furnaces have a per unit capacity of up to 150,000 t/year.

---

[70] Source: Denis & Castor (1994)

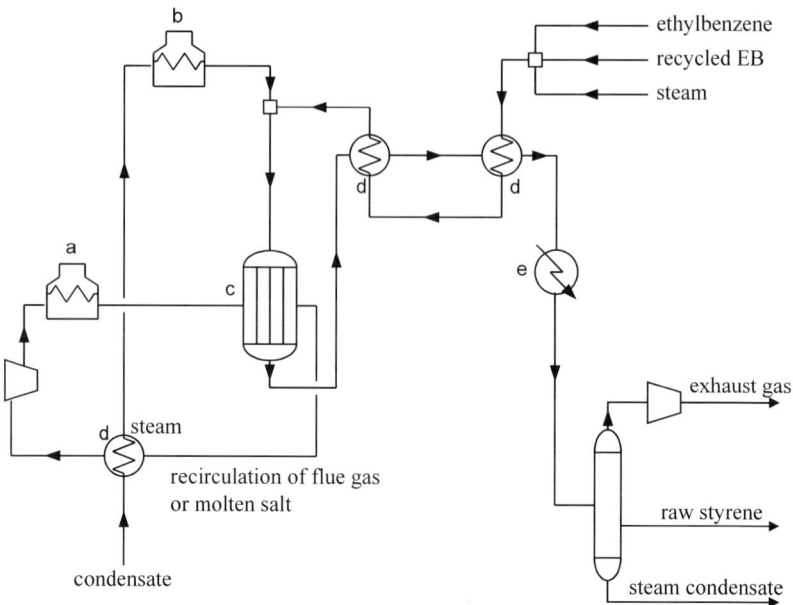

Fig. 24. Process schematic for the isothermal dehydrogenation of ethylbenzene[71]
a) heating; b) steam superheater; c) reactor; d) heat exchanger; e) condenser

The addition of steam is an inherent disadvantage of both procedures, as the energy for producing superheated steam can only be minimally recovered. Another weak point is the reversibility of the dehydrogenation process, which inhibits maximum styrene output. The maximum styrene conversion is therefore limited to 40–50% in the first pass, even in modern plants. However, to ensure sufficient selectivity (and as few side reactions as possible), a multi-stage plant brings the conversion rate up to 65–70%. This means that 30–35% of the original ethylbenzene passes by the reactors unprocessed and must be separated and recovered in an energy-consuming procedure.

A purity of more than 99.8% is required to make the styrene usable for subsequent processes such as polymerisation. Separation of ethylbenzene and other byproducts from styrene, a process that relies on the tendency of styrene to polymerize and the 9°C difference in boiling point between styrene and ethylbenzene, is extremely difficult and very cost-intensive.

---

[71] Source: Denis & Castor (1994)

4.3 Case study 2: Nanotechnology Innovation in Styrene Production

Aspects of catalyst implementation

The catalyst consists primarily of iron oxides (80%), potassium (10%), and chromium oxide, as well as other selectivity-increasing heavy metal promoters (such as Cr, Ca, Al, V, W or Li). The styrene catalyst is classified as an SLP catalyst (supported liquid phase). These catalysts consist of a porous solid carrier, which can be catalytically inert or active. On the carrier we find the catalytically reactive active components (promoters) in the form of a fused material or in a liquefied state (Hagen 1999).

The potassium-promoted iron oxide catalyst is gradually deactivated by use and must be replaced every two to three years. Considering the vast size of the reactors used in the industrial process, it can be assumed to be an extremely cost-intensive process. Besides the already mentioned causes of deactivation of the catalyst - blocking and coke deposit – another three reasons for gradual failure of the catalyst are given (Maximova 2002):
- Loss or redistribution of the potassium promoters
- Major changes in the oxidation state of the iron oxide
- Physical deterioration of the catalyst

One can give rise to the other and all take place simultaneously; the deactivation of the iron oxide catalyst can therefore be viewed as an extraordinarily complex process.

Variant 2: Styrene synthesis with carbon nanotube catalyst

By being able to get a nanometer-scale "look" at events taking place during catalytic styrene synthesis, the actual sequences of the styrene synthesis could be recorded by scientists at the Fritz-Haber-Institute of the Max-Planck-Society. New research suggests that the coke layer is constantly present during styrene synthesis, even on the active catalyst surface. The coal gasification and coke formation are in a balanced state. Therefore it is assumed that coke is not simply a black layer that promotes deactivation, but rather that there are different types of coke with different properties (Ketteler 2002).

These investigations suggest that the carbon film that forms contains the real catalytically active species and that the potassium iron oxide film is only necessary for the formation of this active carbon species, i.e. it serves only as a "co-catalyst." Moreover, it has been demonstrated that various carbon species (carbon black, graphite, "nanobulbs," and "nanofilaments") all show excellent activity and output (MPG 2002).

The enterprise Nanoscape AG was founded in 2001 with the goal of technically implementing the results of this fundamental research. Their aim is to develop a new method of technical styrene synthesis using a nanotube-based catalytic process. Following the successful production of styrene on a small-scale, this is now being expanded as part of an EU project to a reactor with a 100 g catalytic volume. This will be followed by a correspondingly larger pilot plant.

A new nanostructure catalyst consisting of multi-wall nanotubes will be used. This not only permits increases in styrene output, it also changes the procedure from an energy-intensive endothermic process to a more energy-efficient exothermic process. Additionally the new catalyst makes it possible to run the reaction by adding air instead of water. Moreover, at the same conversion rate selectivity can be increased and the process temperature lowered by 200°C, which significantly lowers the specific expenditure of energy.

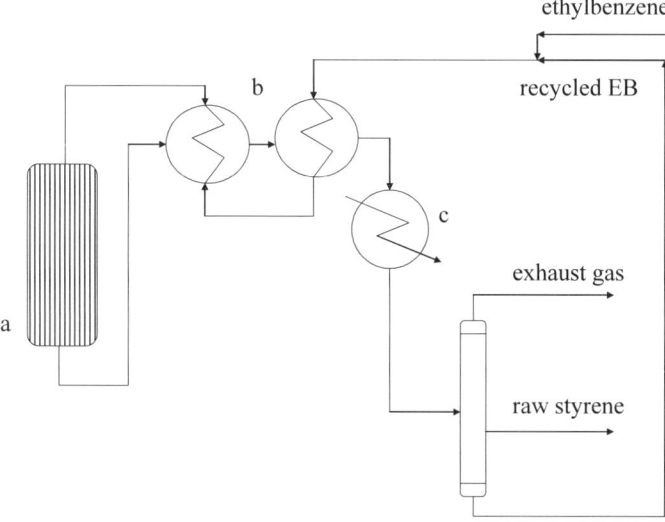

Fig. 25. Process schematic for the oxidative dehydrogenation of ethylbenzene (EB)[72]
a) reactor; b) heat exchanger; c) condenser

The plant schematic makes clear that this process is also characterised by a simpler plant structure as compared to the traditional production of styrene. The advantages of the new styrene production process on the basis of nanotube catalysts and the associated ecological effects can be summarized as follows.

---

[72] Source: Mestl (2004)

4.3 Case study 2: Nanotechnology Innovation in Styrene Production

Table 23. Advantages of the new styrene synthesis using a carbon nanotube catalyst[73]

Advantages of the new procedure	Ecological impact
Change of reaction type from endothermic ($\Delta H^{600°C} = 124.9$ kJ/mol) to exothermic ($\Delta H^{400°C} < 0$ kJ/mol)	Reduction of the specific energy consumption by (at least) 1.2 MJ/kg styrene assuming $\Delta H = 124.9$ kJ/mol, Dependent on the exothermic conditions, which technologically can be kept low; moreover waste heat from the reactor could be used for other processes, such as preheating of reaction gas, heating of tubes, etc.
Reduction of the reaction temperature by about 200°C from 600°C to 400°C	Reduction of the specific energy consumption
Change of the reactive medium from superheated steam to nitrogen/oxygen or air	Reduction of the specific energy consumption, since production and processing of steam is very energy-intensive; reduction of the plant costs, requirements of reactor construction, heat exchanger (e.g. process water separation is eliminated completely), etc., tube dimensions are reduced due to lower temperature level and different corrosion characteristics of the reactive media
Higher selectivity at same conversion rate	Reduction of the specific energy consumption for less distillation and recycling
Use of carbon nanotube catalyst	Replacement for heavy metals, no heavy metal contamination. Easier catalyst management (assumption)
Catalyst production	Higher (energy) expenditures for the production of the nanotube catalyst are to be expected with the CVD process, knowing that the technical requirements for multi-wall nanotubes cannot be compared to those of single-wall nanotubes.

Detailed life cycle assessment data for the alternative styrene synthesis are not available. However, on the basis of the description of the technology, the following deduction about the energy consumption for this process will be made in order to assess the energy consumption at this process stage:

---

[73] Source: Mestl (2004) and authors' own data

1. The change of reaction type results in a reduction of the specific energy consumption by 1.2 MJ/kg styrene.
2. Reduction of the reaction temperature by about 200°C from 600°C to 400°C effects a 25% energy savings.
3. With the change of reaction medium from superheated steam to nitrogen/oxygen or air as well as the higher selectivity at the same conversion rate, a further 5% saving of energy can be assumed.

4.3.4 Life cycle inventory analysis

In the life cycle inventory analysis, the material and energy relationships of the various styrene production processes are noted, with a view to possible environmental impacts, i.e. inputs from the environment and outputs into the environment are recorded. The goal of the life cycle inventory analysis is to establish a data inventory based on functional equivalents for the selected variants.

Life cycle assessment data are available in summary form for traditional styrene production. The data encompass all processes, from crude oil and natural gas extraction and processing to styrene production. Intermediate processes as well as energy and transport processes are also included (APME 1999). As an example, gross energy demands are represented in the following. Additional LCA data are available in the appendix (Table 45ff).

Table 24. Gross energy demand in MJ for the production of 1 kg styrene[74]

Fuel type	Fuel prod'n & delivery energy	Energy content of delevered fuel	Energy use in transport	Feedstock energy	Total energy
Electricity	1.36	0.62	0.01	<0.01	1.98
Oil fuels	0.80	12.12	0.26	16.68	29.86
Other fuels	2.07	19.76	0.05	30.26	52.14
Totals	4.22	32.50	0.32	46.94	83.98

Table 25. Gross material and fuel demand in MJ for the production of 1 kg styrene[75]

Fuel type	Fuel prod'n & delivery energy	Energy content of deleverid	Energy use in transport	Feedstock energy	Total energy

---

[74] Source: (APME 1999)
[75] Source: (APME 1999)

	fuel				
Coal	0.39	1.21	<0.01	<0.01	1.60
Oil	0.69	12.18	0.27	16.68	29.82
Gas	2.49	23.10	0.04	30.26	55.89
Hydro	0.04	0.03	<0.01	-	0.07
Nuclear	0.53	0.25	<0.01	-	0.78
Lignite	0.06	0.03	<0.01	-	0.09
Wood	-	-	-	<0.01	<0.01
Sulphur	-	<0.01	<0.01	<0.01	<0.01
Biomass	<0.01	<0.01	<0.01	<0.01	0.01
Hydrogen	<0.01	0.04	<0.01	-	0.04
Recovered energy	-	-4.34	<0.01	-	-4.34
Unspecified	0.01	<0.01	<0.01	-	0.01
Peat	<0.01	<0.01	<0.01	-	<0.01
Total	4.22	32.50	0.32	46.94	83.98

For further discussion there arises the question as to what portion of these material and energy flows and/or the associated total environmental impact is to be attributed to the styrene process under investigation.

As already discussed, using the available data it is only possible to make a precise differentiation with respect to the energy requirements of styrene production. This reveals that the feedstock proportion, i.e. the energy content of the material, makes up the greatest share (56%) with respect to the gross energy demand. The direct energy demand for styrene production makes up 44% of the gross energy demand at 37.04 MJ/kg.

Table 26. Energy demand in MJ for the production of 1 kg styrene and/or intermediate products[76]

	Benzene production, APME	Ethylene production, APME	Styrene production APME data	Styrene production APME data	Styrene production, Gabi data
Feedstock	45.54	47.73	46.94	46.94	46.94
Raw material extraction	2.83	3.08	4.22	4.22	4.22
Transport	0.21	0.12	0.32	0.32	0.32
Production	19.14	16.10	32.50	17.62	26.45
Part of ethylbenzene + styrene				14.88	

[76] Source: (APME 1999; Gabi 4 Datenbank 1999b; Gabi 4 Datenbank 1999a), authors' calculations

production Part of styrene process					6.36
Sum	67.72	67.03	83.98	83.98	84.29

The actual styrene process requires about 6.4 MJ/kg of the 37.04 MJ/kg. This corresponds with a 17% share of the energy demand for styrene production. This share is influenced by the higher efficiency of the alternative styrene production process.

Table 27. Energy demand for the alternative styrene synthesis

Potential savings in energy	Traditional styrene synthesis	Alternative styrene synthesis
Change of reaction type		- 1.20 MJ/kg
Reduction of reaction temperature		- 1.59 MJ/kg
Change of reaction medium, higher selectivity at the same conversion rate		- 0.32 MJ/kg
Energy demand	6.36 MJ/kg	3.25 MJ/kg

The alternative process based on a nanotube catalyst already yields at this stage a potential energy saving of almost 50%.

4.3 Case study 2: Nanotechnology Innovation in Styrene Production

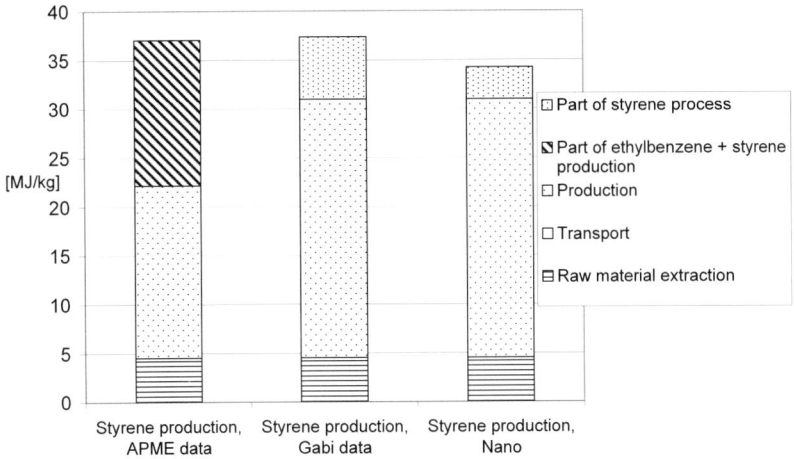

Fig. 26. Comparison of the energy demands of the styrene production processes[77]

With respect to the total energy requirement for styrene production, this would result in an 8–9% increase in energy efficiency.

Another advantage of the alternative styrene production process is the replacement of heavy metals, which would otherwise be present in the catalyst. Heavy metal emissions into water from styrene production would be reduced by about 75%.

---

[77] Source: authors (database: authors', APME 1999; Gabi 4 Datenbank 1999b; Gabi 4 Datenbank 1999a)

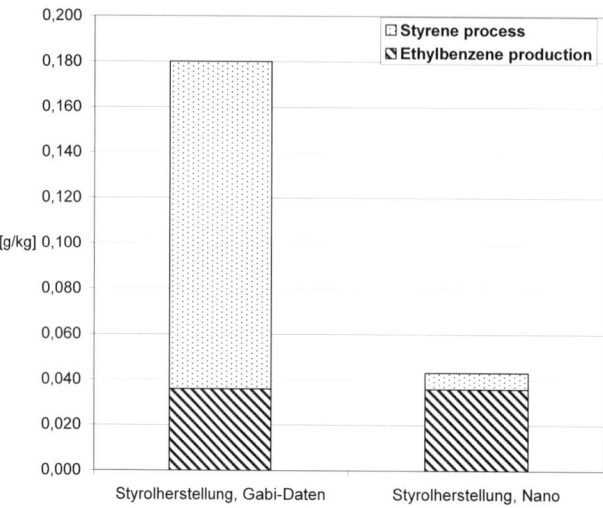

Fig. 27. Comparison of heavy metal emissions into water of the styrene production processes[78]

However, in real-world applications care must also be taken with the new nanotube catalysis method to minimize to the extent possible emissions of carbon nanotubes. The nanotubes should be firmly bonded to the carrier structure of the catalyst so as to avoid "tear off" of the nanotubes from the carrier surface. Moreover nanotube emissions should be minimized by means of sealed-plant technology. A discussion of possible risks and hazards associated with nanotube emissions takes place in the specific case study with a focus on risk potential.

4.3.5 Life cycle impact assessment

For the life cycle impact assessment, it is necessary to have access to emissions data which can be associated with specific environmental impacts. Inasmuch as the available data only allowed for a reasonable calculation of the energy demand, no additional impact assessment was possible.

---

[78] Source: authors (database: authors', Gabi 4 Datenbank 1999b; Gabi 4 Datenbank 1999a)

4.3.6 Impediments to the introduction of nanotechnology to the marketplace

In the context of this case study it makes sense to discuss the subject of impediments to the launch of nanotechnology. It is not necessarily a given, that new nanotechnological processes, for example for styrene synthesis, that demonstrate environmental as well as economical advantages will succeed in the market place.

Within the context of this study, a more indepth interview was conducted to look at the market implementation of this nanotechnological solution. From the results it is clear that a number of styrene producers who have been introduced to this process do not deny its potential advantages. Since it is a procedure which so far has only been realized on a laboratory scale, upscaling up to a larger-sized plant is necessary to prove the economic feasibility of the nanotechnological process. This requires considerable investment in further development, with all the risks that such development entails. The financing for this is not yet settled.

At the same time, we find plant manufacturers and operators using the traditional processes who do not plan any re-investment and who point to their already extensive experience with the existing technology. In the end, it is these sunk costs or unrecoverable past expenditures (already invested capital as well as plant production experience) of the facility operators that may finally prevent implementation of the new process.

4.3.7 Case-study Summary

In the course of this case study, we will investigate and look at an example of the ecological impact of nanotechnology-based catalytic applications. Use of a nanostructured catalyst based on nanotubes for the chemical process of styrene synthesis serves as a specific example.

Implementation of the new catalyst would greatly increase energy efficiency (by almost 50%) at the process level. This improvement in efficiency results from two specific effects: first, it is possible to replace the former endothermic reaction with an exothermic one; second, the reaction temperature can be lowered considerably, the reaction medium altered, and the plant power input minimized. With regard to the overall styrene production life cycle, this would mean an increase in efficiency of about 8–9%. Furthermore, the new catalysis would make possible considerable reductions in heavy metal emissions during the product life cycle. Investigations into potential risks associated with the utilization of nanotubes must continue and be accounted for in facilities planning.

4.4 Case study 3: Nano-innovations in displays

4.4.1 Contents, goals, and methods

The display market is a dynamic and immensely growing market due to the importance of display screens in the fields of information and communication technology. The present sales volume of 26.5 billion dollars in 2005 is expected to reach 49.6 billion dollars in 2005 (Euroforum 2003). Other sources even assume growth to 100 billion dollars in 2005 (Mounier 2002).

Several different display technologies are competing in the market. In the past decade, display technology research and development has intensified with respect to the application of nanomaterials. Formerly CRT displays were the most common in use. These are increasingly being squeezed out of the market by new display technologies. The new display technologies offer many advantages: improved technical specifications (e.g., reduced radiation emissions) and easier handling (e.g., smaller size and less weight). So far, only a slightly weaker performance in some respects (visual angle, brightness, etc.) and, above all, a higher price have kept the new technology from spreading even faster.

In the course of the case study, a comparative investigation of conventional and nanotechnology-based display technologies in the form of a life cycle assessment profile was carried out. Possible eco-efficiency potentials were quantified to the extent possible.

4.4.2 Scope of the investigation

4.4.2.1 Subjects of the investigation

The detailed investigations in this case study are limited to the following display technologies:
- Cathode-ray tube – CRT
- Liquid-crystal display – LCD
- Organic light-emitting display – OLED
- Plasma display panel – PDP
- Carbon nanotube-based field-emitter display – CNT FED

The CRT, the conventional and currently most widely used display technology, and the likewise well-established LCD face two new nanotechnological rivals: the OLED and the CNT FED. These are still primarily in research and development, but some product applications using OLEDs are already on the market. Plasma displays are already available commercially in various configurations and are being promoted as the best solution for large displays and information read-outs. In exactly this segment of the market, FEDs are said to have great potential; comparisons are thus made to plasma technology.

Variant 1: the cathode-ray tube (CRT)
The cathode-ray tube is the oldest and best known facility for generating moving images. A CRT monitor consists of a vacuum-filled (10^{-6} to 10^{-7} torr) glass bulb plus a heated cathode (voltage about 25 kV), also known as the hot cathode or electron gun (Abrams et al. 2003). When heated, the electrons of the negatively charged cathode begin to oscillate and are then emitted from it. Between the cathode and the anode exists an accelerating potential of several kV. Due to this voltage difference, the electrons are accelerated in the direction of the anode and generate a point of light when they strike a phosphor coating on the side of the glass vessel (Tannas 1985). Color is generated by three individual electron beams, which strike differently endowed phosphor layers through a hole or a slotted mask, thus generating red, green, and blue light. The grid regulates the intensity of the electron beam and thus the brightness of the resulting light spot and is controlled by the video brightness signal. The electron beam passes through an electromagnetic field (the deflection yoke) and is thus able to reach every point of the screen. By means of a focusing as well as magnetic coils and a shadow (slot) mask, the electron beam is concentrated and directed to a corresponding pixel (see Lohmann 1997; Blankenbach 1999; and others).

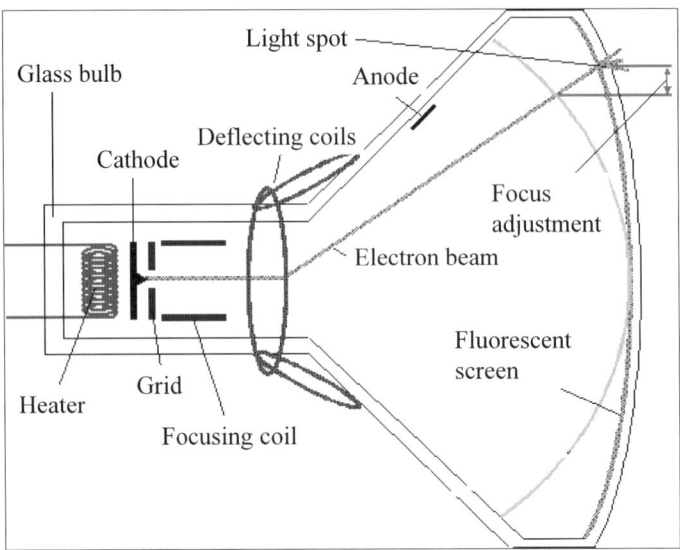

Fig. 28. Schematic diagram of a CRT display

Due to their construction, CRTs are very heavy as compared to other display technologies, require a lot of space, and are characterized by high energy consumption. Image quality is also negatively affected by screen burn-in, which occurs when a fixed image is displayed over a longer period of time. The advantage they offer over other display technologies is their considerably lower price.

Variant 2: the liquid-crystal display (LCD)
Liquid-crystal displays are passive, i.e., transmissive, displays requiring a light source placed behind the screen. The crystals function like a valve, either allowing light to pass or blocking it. This function is based on the anisotropic properties of liquid crystals. Their liquid crystalline aggregate state combines the molecular orientation in the solid, crystalline phase with the mobility of the liquid state.

A liquid-crystal display consists of a system of two glass plates with a metallized lattice of conductors; positioned between them are the liquid crystals. In addition, a polarizer filter is positioned in front of and behind the glass plates, respectively, such that their orientation is crossed. If a light beam is sent through the first polarizer, only one polarization component remains. Unless a voltage is applied, the light is then rotated in its plane of polarization by the liquid crystal and thus can pass through the second polarizer filter, resulting in a bright pixel. When a voltage is applied, the light is no longer refracted and cannot pass through the second filter; a dark pixel is the result (see Chalamala 2000; Theis 2000; Lueder 2001; and others). There have been a number of different developments: passive-matrix, active-matrix LCD (AMLCD), twisted-nematic LCD (TN LCD), ferro-electric liquid crystal (FLCD), and others (Nocula & Olbrich 2003), which are not discussed in this study.

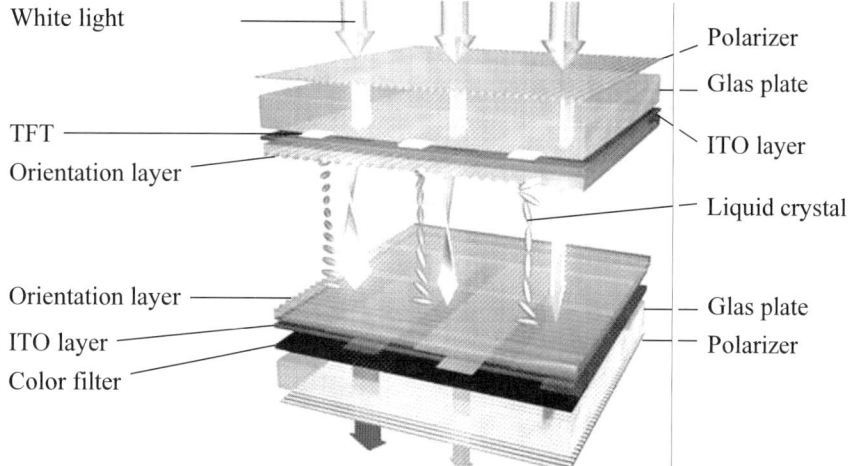

Fig. 29. Schematic of a liquid-crystal display[79]

The greatest advantage of LCD technology is the relatively small size of the displays and their low energy consumption (see Nocula & Olbrich 2003 and others).

Variant 3: the plasma display (PDP)
PDP – plasma display panels – consist of a system with two glass plates, each with a metallized lattice of conductors; between them is a mixture of inert gases such as argon or neon. An electric field is created at the crossover points by applying a voltage to the grid of conductors. This field raises the gas atoms to a higher energy level. When returning to the original level, the absorbed energy is emitted in the form of photons.

Each pixel is directly generated "on site" by its own light source. A large pixel matrix is situated between the flat glass panels. Each pixel consists of three cells, each filled with inert gas. As a rule, neon and xenon are the main components; some manufacturers also add helium. The amount of gas used is very small and the pressure is also minimal. The three cells each have their own wall coating which consists of different phosphor

---

[79] Source: Merck KGaA Darmstadt (2000)

mixtures. Charged electrons create tiny gas explosions which cause short-term changes of the aggregate state from gas to plasma. The resulting ultraviolet radiation generates – depending on the coating of the rear and lateral sides of the cell – red, green and blue light via the phosphors. This phosphorescent light is visible as a pixel through the front panel. The process of light generation is similar to that of the fluorescent tube, but on a much smaller scale, with the result that the energy efficiency of the discharge is only about six percent (Jüstel et al. 2000).

Its advantages include outstanding image quality (fully distortion and flicker-free, high resolution), a wide viewing angle, and immunity from electromagnetic interference (Blankenbach 1999). Disadvantages of PDP technology include, above all, high energy consumption, weight, and difficulties in achieving high brightness and strong contrast at the same time (see Deschamps 2000 and others).

Variant 4: the OLED

OLEDs (organic light-emitting diodes) differ from the usual LED displays through the use of organic emitter materials. The principle functionality of an OLED, like the inorganic LED, is based on injection electro-luminescence. Positive and negative charge carriers, which are injected at the respective electrodes, are brought to radiative recombination in an emitting layer. A DC voltage of only a few volts is sufficient for injection electro-luminescence. A significant advantage of OLEDs is their independence from the substrate material (Scott et al. 2000; Steuber 2000).

Organic light-emitting diodes were invented by C. W. Tang and S. A. Van Slyke (1987) of Kodak, who were the first to discover that organic semiconductors of the p- and n-type could be combined – in a way similar to the formation of p-n-transitions in crystalline semiconductors – to make diodes. Moreover, the polymers generate light by recombination of holes and electrons in a very efficient manner, similar to gallium-arsenide and III-V semiconductors. In contrast to the manufacture of III-V LEDs, where crystalline perfection is necessary, organic semiconductors can be vacuum-metallized as amorphous layers. Two OLED structures are depicted in Fig. 30.

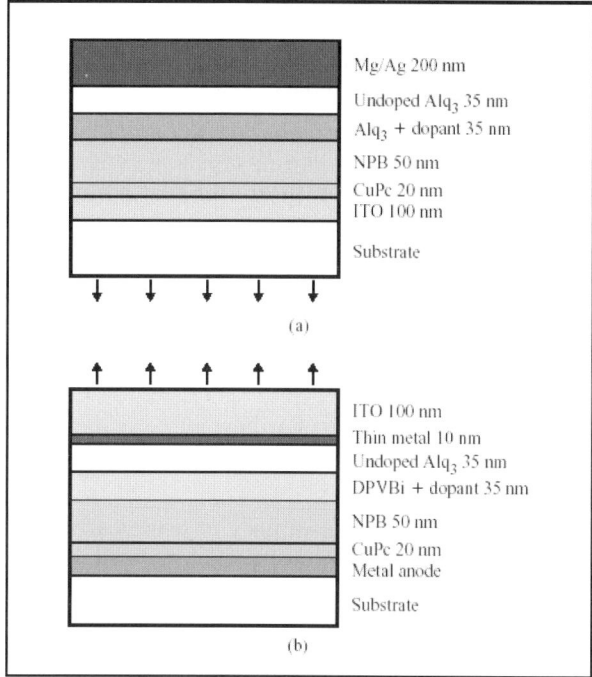

Fig. 30. Schematic structure of an OLED
(a: down-emitting stack (green emission); up-emitting stack (white emission))

By means of such a permutation of layers, a bright emission of green light with a 10–12 cd/A quantum yield and roughly 4 lm/W at 2,000 cd/m² energy efficiency was achieved and utilized in products of the Pioneer Corporation.

T. A. Ali, A. P. Ghosh, and W. E. Howard of IBM and eMagin Corporation have modified this structure to integrate OLEDs on a silicon chip with the intention of using it for micro-displays in headsets. They start with a high work function metal anode and end with a transparent cathode and an ITO layer. Transparency of the metal cathode is achieved by its negligible thickness of only 10nm. They also modified the active light-emitting layer. By using diphenylene-vinylene (DPV) as a blue-green emitter and doping with red colorant, they were able to generate white light (Ali et al. 1999).

The worldwide technological development of OLEDs is very dynamic; in Germany, a group at the Institute of Applied Photophysics at TU Dresden has succeeded in lowering the operating voltage for OLEDs to 3 V and improving the performance efficiency by p-doping of the hole transport layer (HTP) (Blochwitz 2001). Prototypes of OLED displays up to 40" have already been shown by companies such as Samsung (Samsung 2005) and LG Philipps (Pressetext 2004). Driving this development are the anticipated production savings over other displays.

Variant 5: CNT FED

The basic components of a field-emitter display are the rear plate with the cathode layer, a vacuum gap, and the subsequent front plate with an ITO layer on the interior side that serves as the anode.

The cathode layer on the rear plate (emitter layer) is one of the essential components of FE displays. The type, characteristics, and nature of the emitters are decisive for image quality, energy consumption during use, and service life.

There are several known emitter materials. Farthest along is the research and development of microtips made of molybdenum and tungsten and emitter layers made of carbon nanotubes.

Because of their unusual properties, the feasibility of using carbon nanotubes as field emitters was investigated shortly after their invention.

General requirements for field emitters include the following:
– Stability at high current density
– High conductivity
– Low energy loss
– High chemical stability

These requirements are well met by nanotubes. With their very high length-to-diameter ratio, high current density at low voltage, and high thermal and chemical stability, nanotubes were predestined to be well-suited materials as field emitters. Moreover, field emitter layers can be more easily produced using nanotubes than microtips and can be manufactured at a vacuum of 10-8 torr as compared to 10-10 torr for tungsten and molybdenum (Baughman et al. 2002). This permits more cost-effective manufacture of CNT FEDs than of microtips (Information Society Technologies 2003).

The most intricate component of a CNT FED is the cathode layer. It includes a glass substrate upon which a layer of conductor paths is imprinted. Above this layer the "cold cathode" is situated, an emitter layer consisting of nanotubes. Above that there is another electrode layer, the conductive paths of which lie crosswise to the first one. In this way the second orientation of the pixels is defined, which are situated at each of the junctions of the first and second layers. Electrons from the emitter layer can pass through the gaps etched into the second layer, through the vacuum, to the anode layer.

Between the cathode and anode layers is a vacuum and the distance (as a rule, less than 1 mm) is defined by so-called spacers. Opposite each pixel of the cathode layer is a phosphor area of similar size above a transparent conductor material, which functions as anode. The anode layer is situated on a glass substrate.

The rear electrode contact is made by means of a metal layer on the glass substrate (Burden 2001).

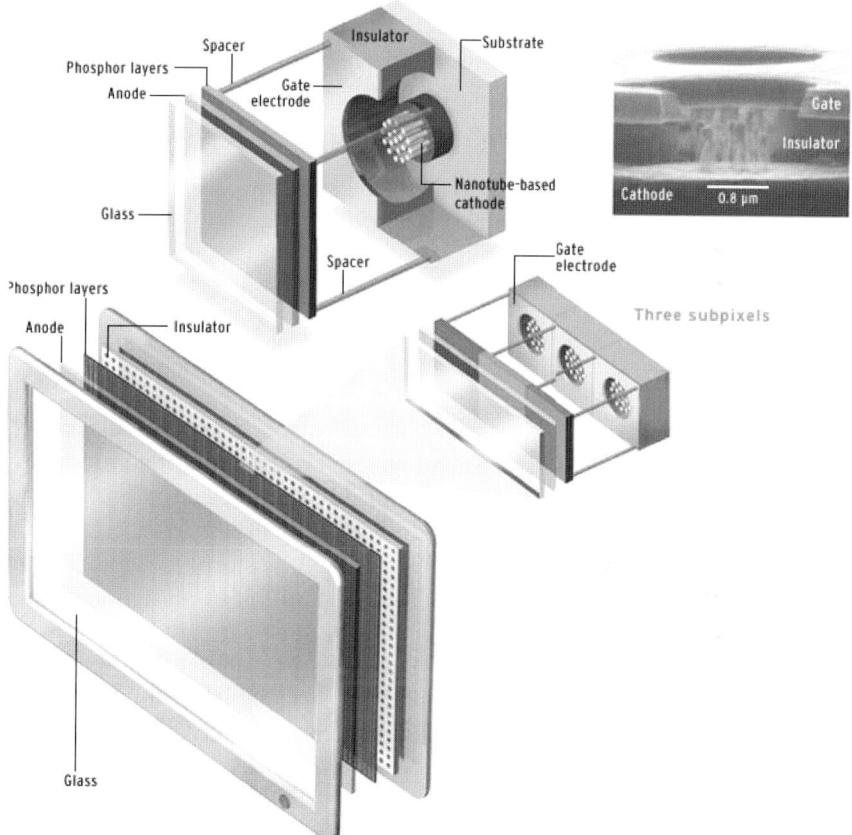

Fig. 31. Schematic layout and enlarged image of a CNT FED[80]

Field-emitter flat screens function much like cathode-ray tubes with many simultaneously emitting cold-field electron sources per pixel and achieve the same performance as the conventional CRT with respect to brightness, color reproduction, viewing angle, and rendering speed.

---

[80] Source: Amaratunga (2003)

During operation, a voltage of 1–7 kV is applied at the anode, depending on the size of the vacuum gap and requirements of the display. Thus a difference in potential (voltage difference) is generated between anode and cathode, thus effecting the emission of electrons from the (nanotube) emitter layer. Acceleration of the electrons in the vacuum gap is thus crucial. Current flow, electron energy, as well as the type and quality of the phosphor determine the color and brightness of the pixels.

A significant advantage of field-emitter displays is that the application of an electric field produces a "cold" electron emission from the tip, with the result that it consumes much less energy than does the traditional cathode-ray tube, which must be considerably heated before it emits electrons.

4.4.2.2 Scope of investigation and availability of data

The scope (system boundaries) of the comparative assessments should cover the entire life cycle of the respective display technologies. The individual life cycle stages include:
- Raw material procurement, pre-production
- Display manufacture
- Use phase
- Disposal/recycling

For the CRT and LCD assessments we were able to rely on a very extensive American life cycle assessment study. The study, commissioned by the EPA Design for the Environment program, was carried out by Socolof, Overly, Kincaid and Geibig of the University of Tennessee, Center for Clean Products and Clean Technologies, and published in December 2001 (Socolof et al. 2001). It is notable for the cooperation of all major American and Asian manufacturers of CRTs and LCDs. The study took as its functional unit for investigation the life cycle of a computer monitor.

No quantitative life cycle assessment data is available for the other variants that were examined. For these technologies, qualitative descriptions of the manufacturing processes were carried out in order to arrive at a basis for making assumptions for the comparative assessments.

With respect to the particularly relevant use phase, quantified data could be provided for all technologies.

Data for the disposal/recycling phase are likewise only available for the first two variants.

Following the example of the existing life cycle assessment, the life cycle of a 15″ LCD monitor and 17″ CRT monitor were selected as the functional unit for the comparative life cycle assessment profile. Particularly for the use phase, the existing data was recalculated for this display size.

For all other variants, the same service life as current CRT and LCD devices is assumed. This specifically makes the assumption that for the OLED the current problem with long-term stability of the organic luminescent material will be solved.

4.4.2.3 Description of the life cycle stages

Numerous raw materials are used in the manufacture of displays. In the following, the essential production steps and raw materials utilized for each of the display technologies considered are described and summarized. Detailed illustrations can be found in the chapter on life cycle stages.

Production of a CRT monitor includes manufacture of the two main components, the panel glass and the glass bell; the glass contains a fair amount of lead. The individual luminescent phosphors are then applied to the panel glass and sealed with a protective coating. A pre-assembled shadow mask or aperture grille is then attached. The two glass components are fitted together to form the glass flask or picture tube into whose neck the cathode is then fused. Subsequently, the air is pumped out and the tube is sealed. After manufacture of the tube, other components such as the deflection unit are added in the final assembly (see Fig. 34).

The following section describes the usual steps in the manufacture of an LCD display. One should be aware that various methods can be used, particularly in applying the liquid crystal layer. First, the rear glass plate with TFT layer and the front glass plate and color filters are produced. After that, the ITO layer is sputtered or printed, followed by application of the hard layer and tempering. Then the polyimide (PI) layers are printed and the cured PI layer is rubbed to enable subsequent orientation of the LC molecules. After that spacers are sprayed on, followed by seal deposition and curing and application of external contacts for later wiring. During the subsequent cell assembly, the two glass plates are aligned and assembled to complete a panel. After curing in the hot press oven, the liquid crystal fluid filling is added to the panels and polarizer filters are applied (Crystec Technology Trading GmbH 2003)(see Fig. 35).

The manufacture of a plasma display begins with the fabrication of two glass plates, usually 3 mm thick, which are subsequently cleaned. Metal electrodes are applied in rows to the front plate. The next step is the preparation of the black matrix. The entire front plate is covered with a transparent dielectric ceramic layer, which cures at a temperature of almost 600°C. The front plate is then covered with a thin layer of magnesium oxide (MgO). On the rear plate, electrodes are also applied (in columns/matrix) and likewise coated with a dielectric ceramic layer. After curing of the ceramic layer, a magnesium oxide layer is applied. Subsequently, the so-called barrier ribs are formed on the parallel electrode surfaces and the phosphors are deposited into the resulting channels. The rear plate is then fired, before the two plates are finally assembled into one panel and fused together. The air is then evacuated from the inner space and replaced by a gas mixture (mostly helium and xenon) at a pressure of about 500 torr (Deschamps 2000)(see Fig. 36).

The manufacture of OLEDs begins with the application of the transparent ITO layer to a glass substrate. In the following steps, all additional organic and metallic layers are then applied by means of thermal evaporation. The organic materials require a temperature of 300–500°C for evaporation, but silver requires a temperature of 1200°C. The layers are applied in a precisely followed series of complex process steps within a vacuum. Subsequently, the front plate is mounted and the entire assembly is sealed air-tight with a form of epoxy resin.

Presently, only one production plant capable of series production of the first OLEDs is in operation, at SK Display Corporation, a joint venture of Kodak and Sanyo (Webelsiep 2003). A prototype of an in-line OLED production system, funded by the Federal Ministry of Education and Research (BMBF), exists at the Fraunhofer IPMS in Dresden (IPMS 2003). The plant consists of eleven process modules, in which 300 x 400 mm samples are coated. The substrates move vertically through the deposition chambers. Up to twelve line sources are available for deposition of the organic layer systems. Additionally, two PVD and inorganic evaporation sources are integrated into each of the electrode deposition systems (see Fig. 37).

The following two illustrations make clear the reduction in manufacturing complexity of the OLED as compared to the LCD (described above).

4.4 Case study 3: Nano-innovations in displays   123

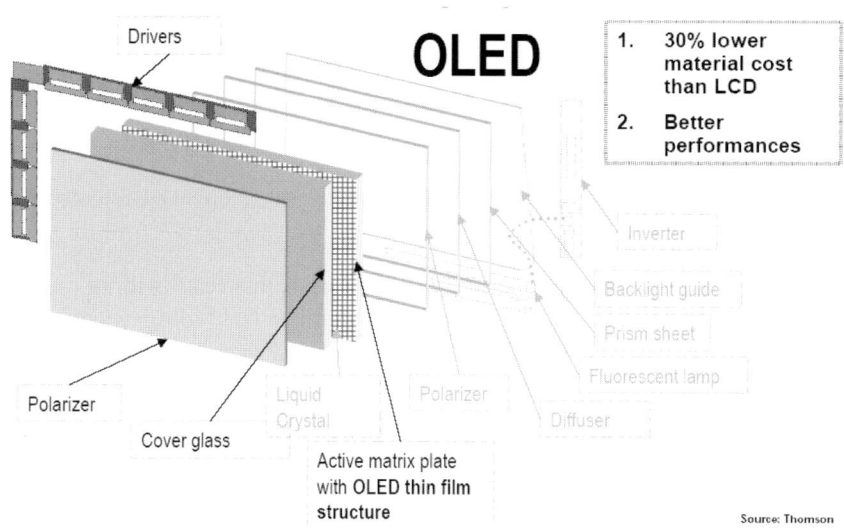

Fig. 32. Comparison of LCD and OLED (1)

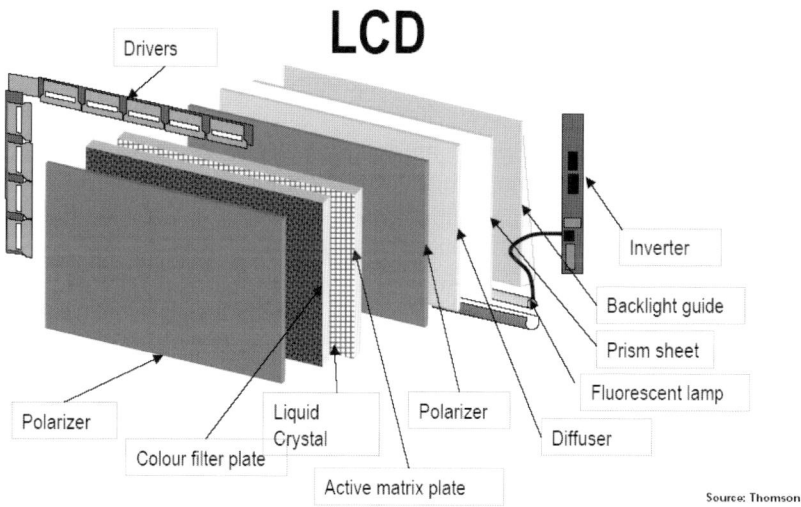

Fig. 33. Comparison of LCD and OLED (2)

From the diagrams it can be seen that the OLED can be manufactured much more efficiently than the LCD because, e.g., no backlight is involved. This makes it possible to reduce the depth of the screen from 5.5 mm in the LCD to 1.8 mm in the OLED.

An assessment of the materials required for a typical OLED:
- Glass (0.7 mm)
- 100 nm ITO
- 100 nm organic material
- 100 nm Mg:Ag (for the contacts)
- Glass (0.7 mm)

for a 17" display with a viewable area of 918 cm², this results in an organic materials consumption of ca. 370 mg per display (including waste). The annual production of 1 million 17" displays would require roughly 0.4 t of organic materials.

Table 28. Estimate of typical OLED materials consumption

	Contained in the display		Total consumption incl. manufacture	
ITO	12 µg/cm²	11 mg for 17"	25 µg/cm²	23 mg for 17"
Organic materials	16–20 µg/cm²	15–18 mg for 17"	400 µg/cm²	370 mg for 17"
Mg, Ag	100 µg/cm²	92 mg for 17"	ca. 200 µg/cm²	184 mg for 17"

After production of the rear glass plate of the CNT FED, metal catalysts are applied to the pixel areas of the glass substrate by means of various procedures, including sputtering, lithography, microcontact printing, and ink jet printing. The carefully controlled growth of the CNT on the catalysts takes place in a precisely moderated CVD process at low temperatures (temperatures less than 500°C and lasting only for a few minutes) (NEDO). On the front plate, which also consists of glass, an ITO layer is applied as an anode to the interior face and is then treated with a phosphor layer. Finally, the two plates are combined to form a panel and bonded together (see Fig. 38).

4.4 Case study 3: Nano-innovations in displays 125

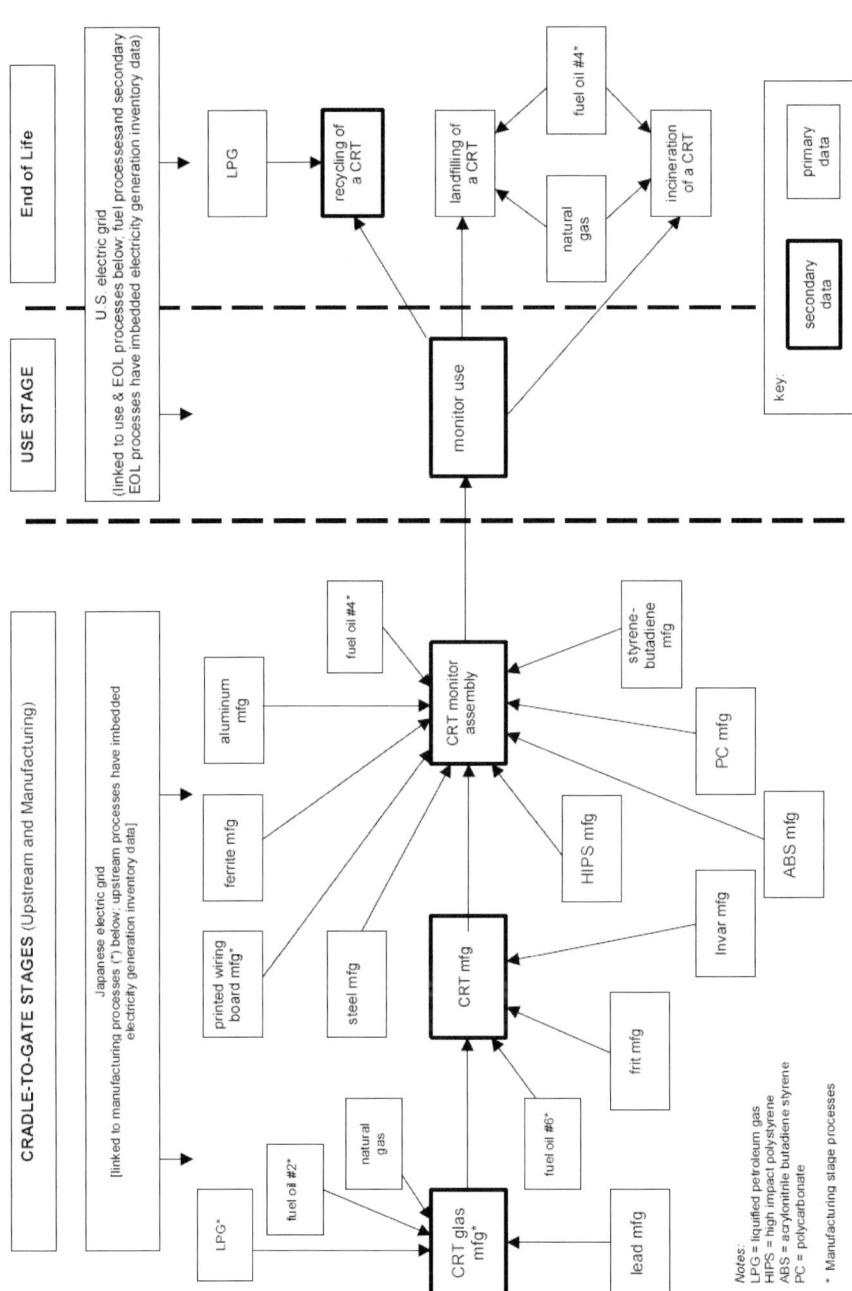

Fig. 34. Product life cycle of a CRT monitor (Source: Socolof et al. (2001))

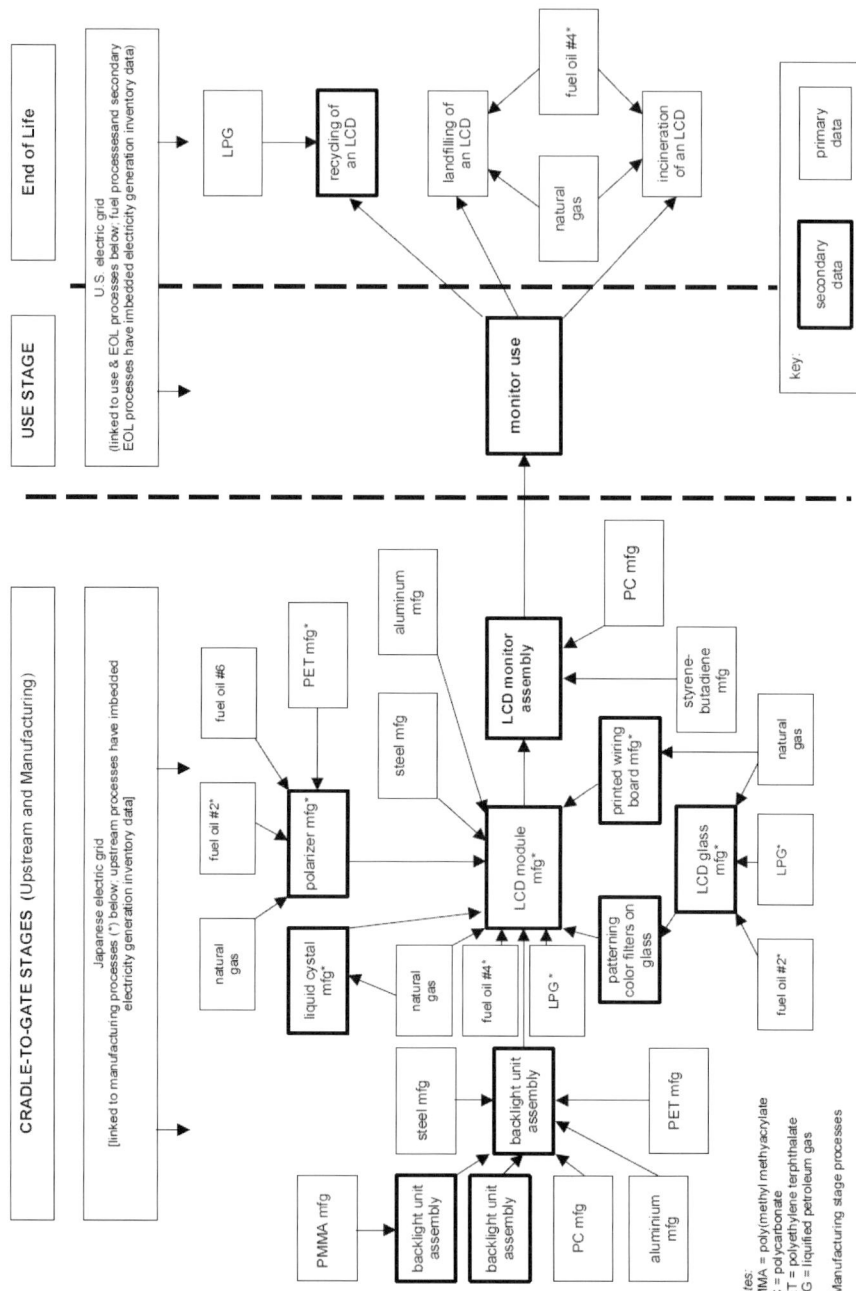

Fig. 35. Product life cycle of an LCD monitor (Source: Socolof et al. (2001))

4.4 Case study 3: Nano-innovations in displays 127

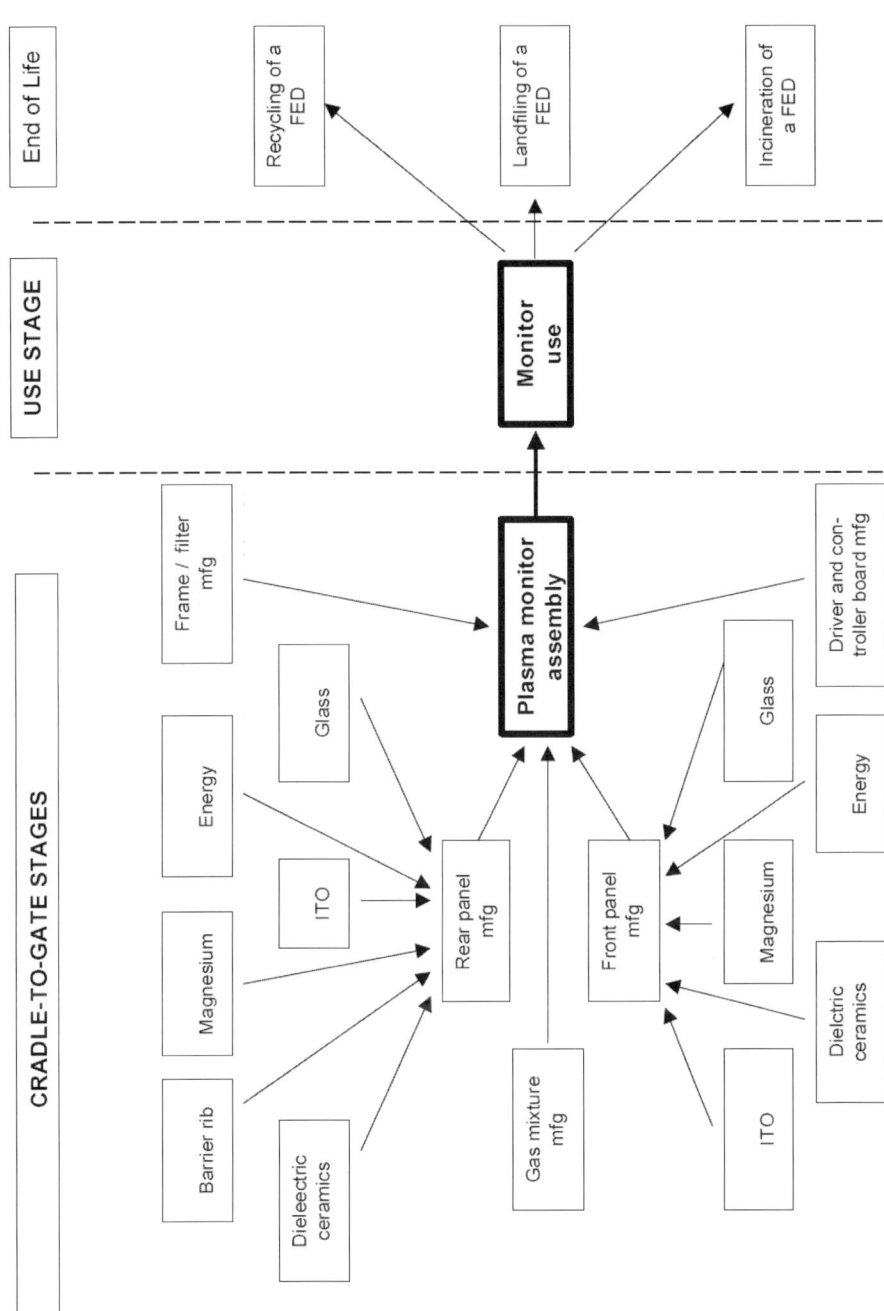

Fig. 36. Product life cycle of a plasma display (Source: authors)

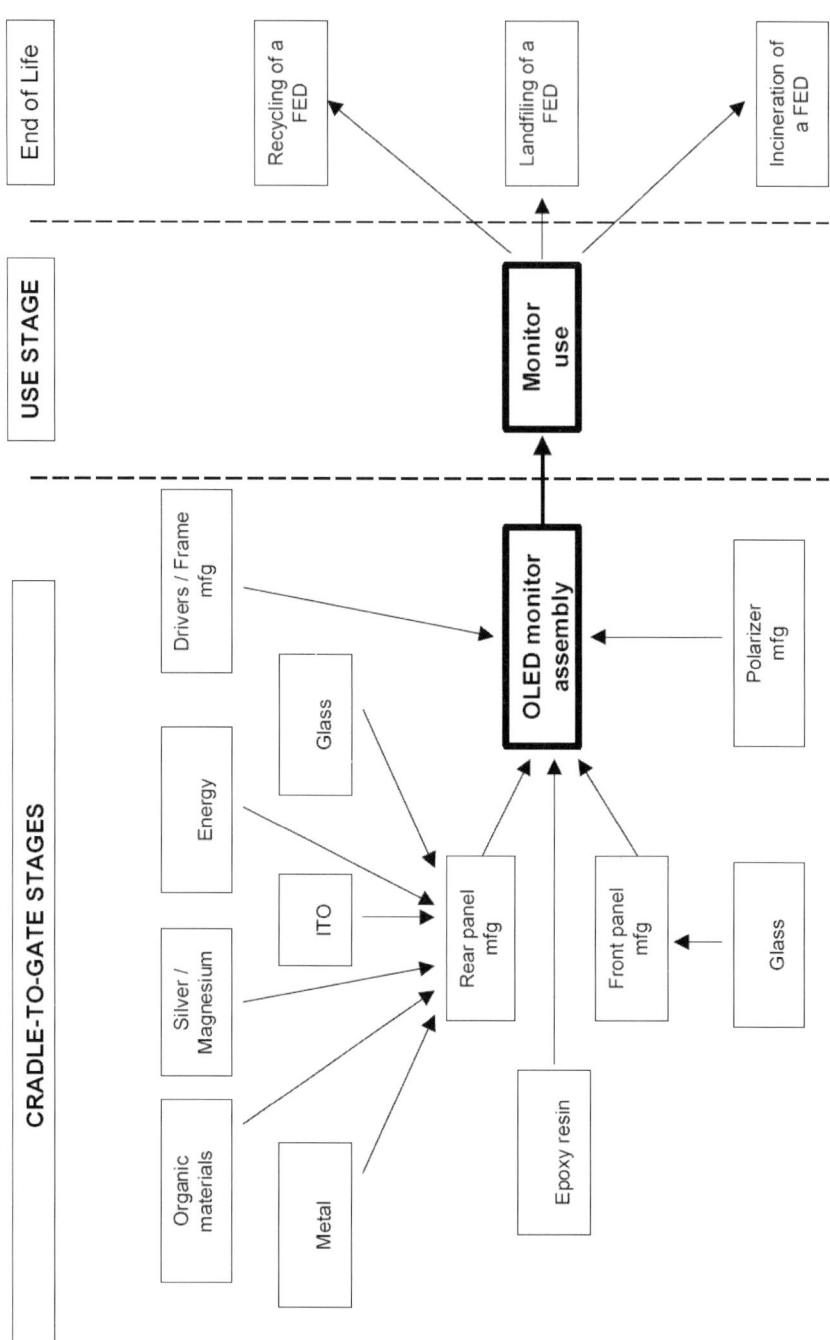

Fig. 37. Product life cycle of an OLED display (Source: authors)

4.4 Case study 3: Nano-innovations in displays 129

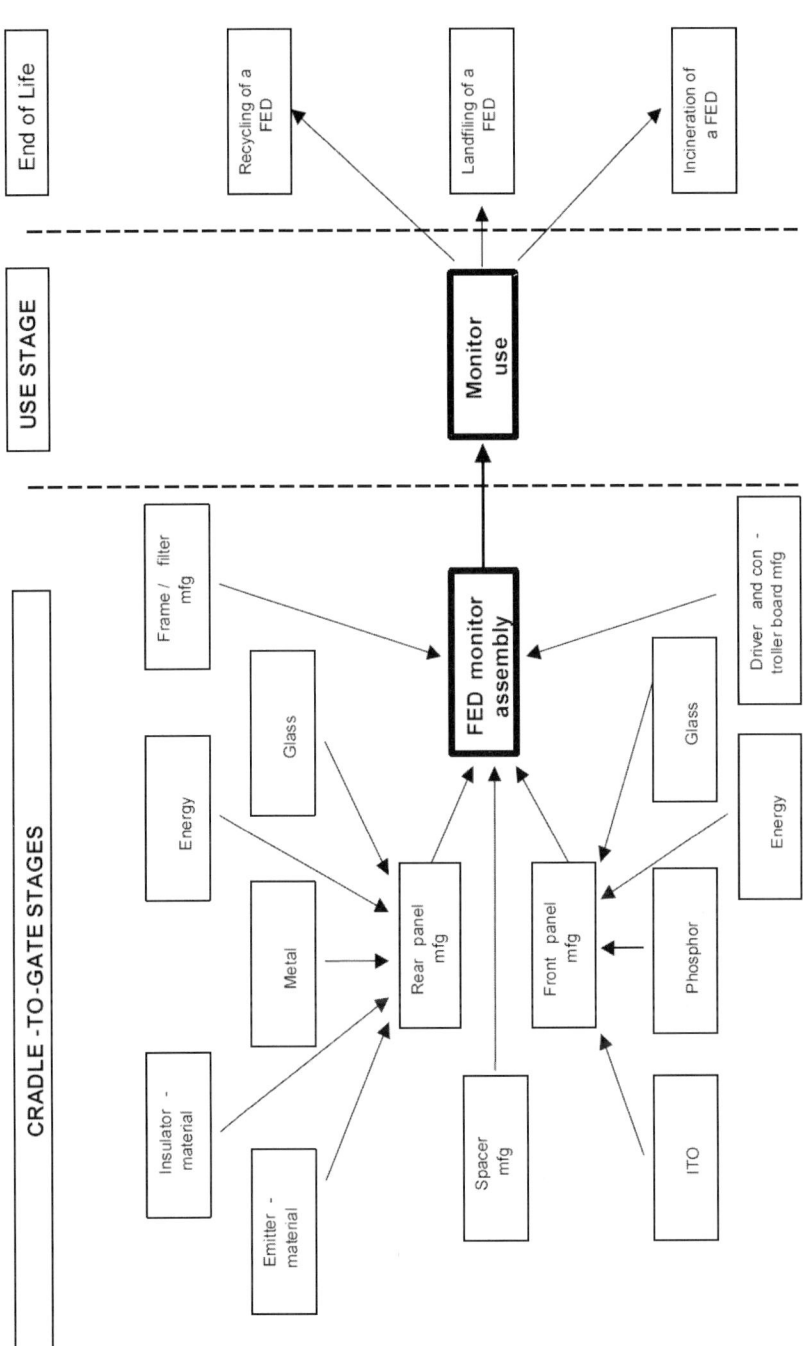

Fig. 38. Product life cycle of a CNT FED (Source: authors)

The advantages of the new nanotechnology-based display technologies and their associated environmental impact may be summarized as follows.

Table 29. Advantages of the new nanotechnology-based display technologies

Display technology	Environmental impact
OLED	
Simple assembly, reduced number of production steps	Reduced facility expenditures
Minimal screen thickness (1.8 mm instead of 5.5 mm TFT LCD), minimal material consumption	Improved material efficiency
Low vaporizing temperature of the organic molecules (300–500°C) as compared to metals (Ag 1200°C)	Reduction of energy consumption during production
Organic phosphors	Problem of long-term stability (shorter service life)?
Reduction of the specific energy consumption in the use phase	Increased energy efficiency
CNT FED	
Reduction of the specific energy consumption in the use phase	Increased energy efficiency
Minimal screen thickness (3.5 mm), less material use	Improved material efficiency
Very high resolution	
Use of carbon nanotubes	Production process complicated by selected partial growth of nanotubes

Detailed life cycle assessment data for the production and prior processes are only available for the CRT and LCD. For the other display technologies examined, the necessary processes could only be qualitatively described, as no material or energy data was available. A differentiated comparative assessment is not possible on the basis of these data.

However, based on descriptions of the technology, the following assumptions could be made regarding energy consumption, pre-production, and production for the three variants, in order to at least be able to make an estimate of energy consumption:

- PDP: The same energy consumption as in the LCD is assumed.
- OLED: On the basis of the simpler assembly, lower energy consumption can be assumed (variant A: 10% energy reduction, variant B: 30% energy reduction).
- CNT FED: On the basis of the simpler assembly, lower energy consumption can be estimated for pre-production (Assumption: As with the

OLED variant A, 10% energy reduction). Also the energy expenditure for manufacturing – even if one assumes application of the most demanding production technology – is likely to be less than for the LCD; however, in order to allow for a safety margin for nanotube production, the same expenditure as for the LCD is assumed.

Use phase

As important as image quality and usability are from a technical point of view, because of their long use phase, electrical energy consumption is the decisive parameter for an environmental assessment of display technology.

To be able to compare our display technologies, we need one (common) energy consumption value for one (common) display size. In the case of the CRT and LCD displays, this was the 15″ monitor size. The energy consumption data available in the literature is for various display sizes, which makes comparisons difficult at best. Cambridge University's Department of Engineering assumes an energy consumption of 50–70 W for a 38″ CNT FED with an image quality comparable to a CRT or plasma display (Amaratunga 2003). Samsung estimates 150 W energy consumption for a 42″ CNT FED (Samsung 2003). Futaba produces an 8″ full-color display with a 1000 cd/m² brightness, which consumes only 7 W. According to Mounier (2002), a 10″ FED display will only consume 2 W. A new 20″ OLED display by IDTech (IBM) with 1280 x 768 pixels uses 25 W at a brightness of 300–500 cd/m². In comparison, the energy consumption of a 19″ LCD display is 40 W (VDI-Nachrichten 2003).

Energy performance data used in the case study are listed in the following table and were taken from a display technologies road map, details of which can be found in the appendix (see Table 51). Data for the year 2005 were converted to the respective display size in the case study as necessary.

Table 30. Energy consumption of display technologies[81]

Year	Technology	2000	2005	2010
Size		32″	50″	50″
Energy-consumption	LCD	140 W	120 W	100 W
	PDP	300 W	200 W	100 W
	CNT FED	70 W	70 W	40 W
	OLED	no data	60 W	30 W
	CRT	200 W	230 W (HVTV)	200 (HVTV)

---

[81] Source: New Energy and Industrial Technology Development Organization (2000)

Disposal / Recycling

Hard data for this life cycle stage for the CRT and LCD displays are found in the life cycle assessment by Socolof et al. (2001). No quantitative data is available for the other display technologies. One can see from the assessment data for the CRT and LCD that this phase plays a minor role in the overall assessment and is not quantitatively addressed in the rest of the process.

4.4.3 Life cycle inventory analysis

In the life cycle inventory analysis, the material and energy relationships between the display technology being studied and the environment are recorded, i.e., the input flows from the environment and the output flows that are returned to the environment are noted. The goal of the life cycle inventory analysis is to establish a data inventory based on functional equivalents for the selected variants.

Complete quantitative life cycle inventory data are available for the CRT and LCD. On this basis and using assumptions described below, an assessment of the energy consumption of all display technologies for the pre-production, production, and use phases was made.

A simple comparison of the total mass of the major components of a 17" CRT and a 15" LCD monitor makes obvious the considerable difference in mass between traditional tube technology and flat screen technology – and that includes the other displays.

Table 31. Composition by mass of CRT and LCD per monitor[82]

	CRT		LCD	
Component	Mass in kg	Share in %	Mass in kg	Share in %
Glass (lead glass in CRT)	9.76	46.1	0.59	10.3
Steel	5.16	24.4	2.53	44.1
Plastics,	3.04	14.4	1.78	31.0
Misc.	3.20	15.1	0.83	14.6
Total	21.16	100.0	5.73	100.0

A look at the bottom line for material and energy consumption over the entire service life underscores the advantages of flatscreen technology. As an example, the total energy consumption of a CRT display is 7.3 times higher than the total energy consumption of the respective LCD display,

---

[82] Source: Socolof et al. (2001)

due to the increased amount of energy needed for glass production. Detailed data regarding material, chemical substances, and energy totals from the life cycle assessment utilized can be found in the appendix (Table 52ff).

The environmental advantages of the LCD versus the CRT become clear, particularly in the comparative list of environmental categories. With the exception of two environmental impact categories, the LCD shows better results than the CRT.

Table 32. Quantitative comparison of environmental categories of the CRT and LCD[83]

Impact category	Units per monitor	CRT	LCD
Renewable resource use	Kg	1.31E+04	2.80E+03
Nonrenewable resource use	Kg	6.68E+02	3.64E+02
Energy use	m3	2.08E+04	2.84E+03
Solid waste landfill use	m3	1.67E-01	5.43E-02
Hazardous waste landfill use	m3	1.68E-02	3.61E-03
Radioactive waste landfill use	m3	1.81E-04	9.22E-05
Global warming	kg-CO2 equivalents	6.95E+02	5.93E+02
Ozone depletion	kg-CFC-11 equivalents	2.05E-05	1.37E-05
Photochemical smog	kg-ethene equivalents	1.71E-01	1.41E-01
Acidification	kg-SO2 equivalents	5.25E+00	2.96E+00
Air particulates	Kg	3.01E-01	1.15E-01
Water eutrophication	kg-phosphate equivalents	4.82E-02	4.96E-02
Water quality, BOD	Kg	1.95E-01	2.83E-02
Water quality, TSS	Kg	8.74E-01	6.15E-02
Radioactivity	Bq	3.85E+07	1.22E+07
Chronic health effects, occupational	tox-kg	9.34E+02	6.96E+02
Chronic health effects, public	tox-kg	1.98E+03	9.02E+02
Aesthetics (odor)	m3	7.58E+06	5.04E+06
Aquatic toxicity	tox-kg	2.25E-01	5.19E+00
Terrestrial toxicity	tox-kg	1.97E+03	8.94E+02

---

[83] Source: Socolof et al. (2001)

What further potential savings or improvements can these other display technologies offer – particularly as compared to those based on nanotechnology? This is a question for further discussion.

Inasmuch as quantitative analyses with respect to substance inputs are not available, eco-efficiency potentials in the area of energy consumption can only be estimated. To do so, existing CRT and LCD data, as well as assumptions derived from them, for the product life cycle stages of pre-production, manufacturing, and use were used to arrive at energy consumption values.

Table 33. Energy consumption of the display technologies in the individual life cycle stages per monitor[84]

Technology	Preproduction [MJ]	Production [MJ]	Use [MJ]
CRT	366	18300	2290
LCD	633	1440	853
PDP	633	1440	1422
OLED 10%	570	1296	427
OLED 30%	443	1008	427
CNT FED	570	1440	498

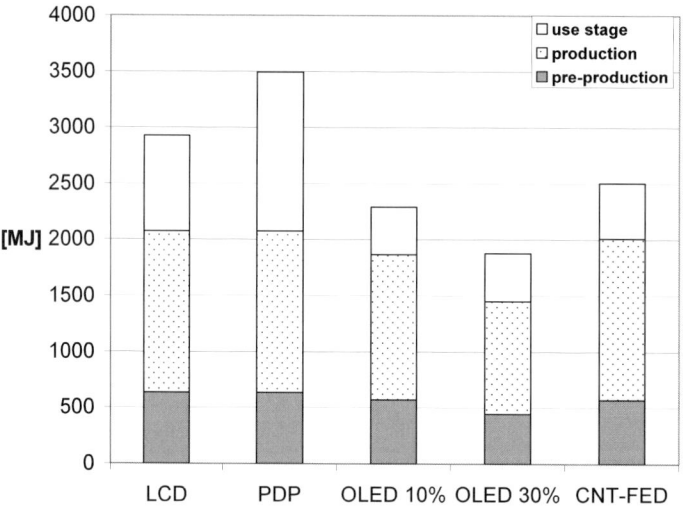

Fig. 39. Energy consumption of the display technologies in the individual life cycle stages per monitor[85]

---

[84] Source: Socolof et al. (2001), authors' own calculations

The energy consumption chart makes clear that the OLED variants, due to lower energy consumption in the use and production phases, fare particularly well when compared to current LCD displays.

Energy efficiency increases and thus an improved eco-efficiency of ca. 20% for the OLED 10% variant and 35% for the OLED 30% variant over the entire product life cycle are possible in this field. Of course, this is dependent on R&D success in achieving the envisaged material and energy efficiencies, as well as solving the problem of long-term stability of luminescent substances.

Under these assumptions, the CNT FED display would also score significantly better as compared to the LCD due to its greater efficiency in the use phase. The eco-efficiency potentials are very much dependant on the CNT FED production process being made as efficient as LCD production.

With respect to difficult-to-assess substances, the study by Socolof et al. (2001) looks at lead, mercury, and liquid crystals. A considerable amount of lead is present in the glass used in the CRT, as well as in the frit, although the lead in the glass is firmly bound in the glass matrix. Mercury is required in small quantities for the LCD backlight (3.99 mg). The mercury is vaporized by application of a voltage and generates ultraviolet light. During the process of manufacturing the backlight there is the risk of mercury emissions due to lamp breakage, mercury leakage, and waste. The risk can be minimized by process optimization and protective measures, but cannot be fully eliminated. Moreover, mercury emissions are also a component in the production of electric energy, as was already discussed in the case study on lighting applications. For the LCD display, an electrical energy production contribution of 3.22 mg mercury was assigned; for the CRT, due to its higher power consumption, 7.75 mg. Similar mercury emissions will occur with the other display technologies.

Due to the lack of sufficient information, no final assessment on the toxicity of the liquid crystals could be made. Toxicological tests done by liquid-crystal manufacturers showed, e.g., that in 95.6% (562 out of 588) of the liquid crystals tested, a potential toxic hazard did not exist. In 99.9% (614 out of 615) of the liquid crystals tested, it could be shown that no carcinogenic risk exists. The German Federal Environment Agency also came to the conclusion that liquid-crystal substances from the company Merck represent a very low risk and that no special requirements are necessary for the disposal of LCDs (Merck KGaA 2000). Instead of liquid crystals, other organic materials are used in OLEDs. Potential risks from extremely dangerous substances in OLEDs are unlikely. The safety data sheets provided

---

[85] Source: authors (Database: authors, Socolof et al. 2001)

to us by the Fraunhofer IPMS regarding some of the substances used in OLEDs do not indicate any potential hazards.

The carbon nanotubes used as field emitters in CNT FEDs are grown on the substrate in a carefully controlled CVD process; the result is tightly sealed in the product. Therefore the potential for the release of carbon nanotubes appears to be very low.

4.4.4 Life cycle impact assessment

To complete the impact assessment, it is necessary to have access to emissions data which can be allocated to specific environmental impacts. Since these are estimates and furthermore a direct relationship between energy consumption and the relevant emissions exists, this would also be seen in the categories greenhouse effect, acidification, and eutrophication and not provide any new knowledge. We therefore make no presentation with respect to this.

4.4.5 Case-study Summary

The objective of this case study was an investigation of the eco-efficiency potential of new nanotechnological products currently being developed for the display industry. For this purpose, OLED and CNT FED displays were compared with conventional CRT, LCD, and plasma displays.

Due to differing stages of development of the technologies under investigation, the resulting eco-efficiency potential assessments come with a certain degree of uncertainty. In the overall product life cycle, the manufacturing phase is responsible for an ever increasing share of the environmental impacts. The successful implementation in mass production of the material and energy efficiency increases offered by OLEDs will make it possible to realize significant eco-efficiency potentials. At the very least, a 20% savings in energy as compared to LCDs over the entire product life cycle should be possible.

Likewise, development of the eco-efficiency potential of the CNT FED will become possible once the manufacturing process, particularly in the highly complex production of nanotubes for field emitters, becomes as efficient as current processes. Risk potentials from these new technologies are unlikely.

4.5 Case study 4: Nano-applications in the lighting industry

4.5.1 Contents, goals, and methods

Light in its many applications is a large consumer of energy. In Germany, about ten percent of the electrical energy consumed is used for lighting (ZVEI 2003). In this case study we investigate the ecological potential of new nanotechnology-based solutions for lighting. The German agenda "Optical Technologies for the 21st Century" identifies light-emitting diodes as an efficient and environmentally favorable light source and is promoting them in the course of its "Optical Technologies" development program under the title "Nanolux – white light-emitting diodes for lighting." Specifically, this case study investigates white LEDs (light-emitting diodes) and compares them with conventional light sources (the incandescent and the compact fluorescent lamp). Evaluation of the ecological relevance is carried out by means of a comparative life cycle assessment. Quantum dots and their potential for improvements in efficiency in the lighting industry are also qualitatively addressed.

4.5.2 Scope of the investigation

4.5.2.1 Introduction

Light can be defined as electromagnetic waves with frequencies in the visible range and thus perceivable as having a particular brightness and color. Waves of other frequencies have names that primarily characterize their use, but no color. Many of our present-day light sources are thermal radiators. This includes the Sun, candles, incandescent, and tungsten-halogen lamps. The luminous color of the object is dependent upon its temperature; i.e. such light sources generate light as a secondary product of heating up.

The other group of light sources generates light by electric radiation, luminescence, or crystal radiation. This includes discharge-type lamps (including compact fluorescents) and the even more advanced light-emitting diodes, which are rapidly and continually being developed in order to make them competitive on the mass market.

138   4Assessment of sustainability effects in the context of specific applications

The following diagram provides a survey of the most important lamp types today. Other discharge-type lamps include the low-pressure sodium-vapor lamp, ultraviolet, sodium-xenon, xenon, reflector lamps, various automotive lamps, etc.

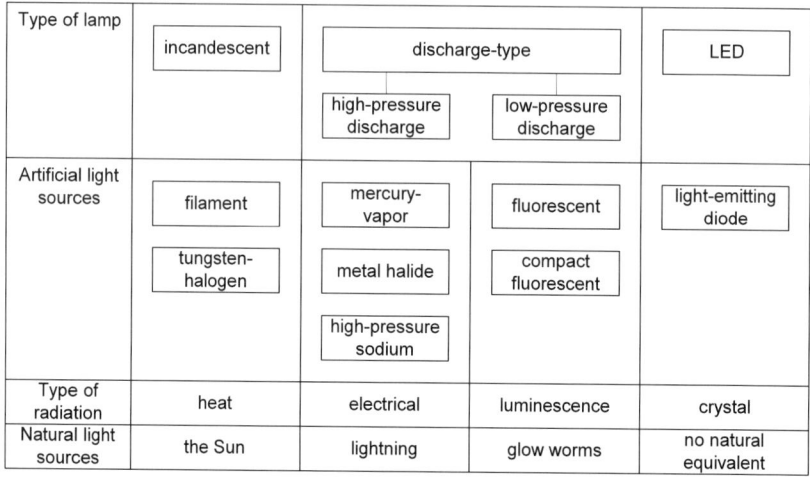

Fig. 40. Overview of the most important types of lamps[86]

4.5.2.2 Subject of the investigation

The case study looked at the use of light sources for illumination. For this purpose, three different types were compared: The two traditional light sources, the incandescent lamp and the energy-saving lamp, and the white LED based on nanoscale layers.

Variant 1: the incandescent lamp

The incandescent lamp, with a luminous efficiency of ca. 15 lm/W (WKO 2003), is the most widely used electric light source. It can be arbitrarily switched on and off and is used in all fields of interior and exterior lighting. The average service life is roughly 1,000–1,500 hours. The incandescent filament, a double-coiled wire between two lead-in electrodes inside a glass bulb, begins to glow when an electrical current is applied. A gas mixture inside the glass bulb prevents rapid vaporization of the filament. Tungsten, with its high melting point of roughly 3,400°C, is used for the

---

[86] Source: Following Grezcmiel (2001)

filament, as it provides a longer service life than would other metals (Wuelfert 2000).

When the current passes through the bulb, the filament is rapidly heated up to a temperature of about 2,600°C. This causes bright light to be radiated. The electrical energy is converted into light very inefficiently – 90–95% of the energy is converted into undesired heat. The incandescent lamp has a very poor energy efficiency. Only a small percentage of the energy is converted into visible light. The energy efficiency can be enhanced by increasing the filament temperature; this requires filling the glass bulb with a halogen mixture to maintain the service life of the lamp. This further development is called the tungsten-halogen lamp. In halogen-filled lamps, the tungsten wire may achieve a temperature of roughly 3,000°C.

Variant 2: the energy-saving lamp (compact fluorescent)
Saving energy and resources is a guiding principle of our time. Using discharge-type lamps allows us to apply energy much more efficiently for lighting than in the case of conventional incandescent lamps. They include the common fluorescent tube lamp, often found in the home and workplace. In particular, energy-saving lamps (compact fluorescents) – a miniature, enhanced form of the fluorescent lamp – make four to five times better use of electrical energy than do incandescent lamps. They are more expensive than normal incandescents, but convert up to 25% of the energy applied to light and have a luminous efficiency of about 60 lm/W (WKO 2003). Moreover, compact fluorescent lamps have a service life of 8,000 to 14,000 hours (KEVAG 2003).

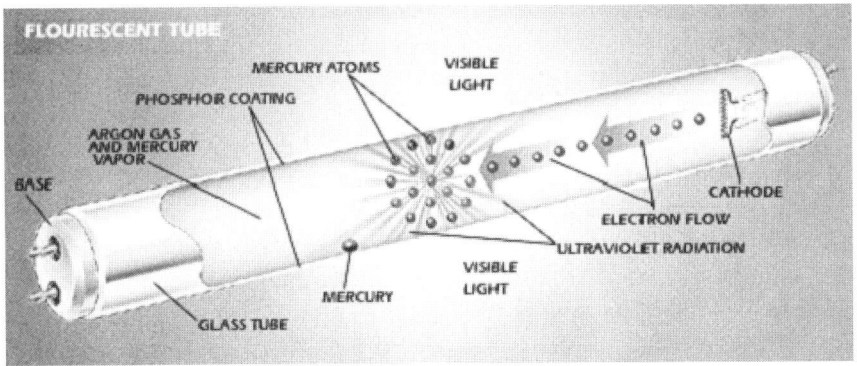

Fig. 41. Structure of a fluorescent lamp[87]

---

[87] Source: Popular Mechanics (2002)

A gas mixture, e.g. mercury, is a major component in discharge-type lamps. When, as a consequence of an applied voltage, a current begins to flow, electrons move from one electrode to the other. The electrons collide with the mercury atoms in the gas mixture inside the tube. These collisions cause energy in the form of ultraviolet light to be released. The ultraviolet light is absorbed by the phosphor coating (e.g. metallic salts) on the inner surface of the glass tube. The phosphors are thus stimulated and emit visible light. A discharge-type lamp requires a ballast, which provides the necessary high-voltage for ignition and also the normal operating voltage.

Variant 3: the (white) light-emitting diode
Recently the use of white light-emitting diodes (LEDs) has come into discussion as an alternative to conventional light sources such as incandescents and fluorescent tubes; this is because it is generally assumed that LEDs can more efficiently produce light. Light generation by means of light-emitting diodes is based on semiconductor lighting technology. This makes it possible, for the first time, to generate a cold light without a significant heat component. The technical problem is that although the internal quantum efficiency of an LED is very high, only a small portion of the light can be decoupled from the component. Internally, up to 90% of the energy can be transformed into visible light, but only about 20–30% (this number is increasing) is externally available (CJ-Light GmbH 2003).

Light-emitting diodes have a very long service life. The service life of the LED is characterized by the number of broken components during a given period and by the decrease of light flux as compared to the original value (50%). A complete failure of all components during service life is practically impossible when operating norms are observed. Service life, efficiency, and light color of the LED depend very much on temperature, with the consequence that the thermal balance must be carefully observed when using LEDs. Additionally, the light-emitting diode can be destroyed by application of a too-high voltage. Additional electronic components in the form of a ballast are necessary for LED systems and modules.

The maximum service life of 100,000 hours can be achieved in red, yellow, and orange LEDs under normal ambient temperatures and at 50% of the maximal allowed current recommended by the manufacturer. The service life of white light-emitting diodes is greatest at an ambient temperature 30 K below the specified value and at 50% of the original current rating (FGL 2003). White LEDs have an average service life of about 15,000 hours.

At the core of light-emitting diodes are semiconductor crystals that generate light when a current is sent through them. Inside the crystals is an n-conducting region having a surplus of electrons (negative charge) and a p-conducting region with a deficit of electrons (positive charge). Between those is a transition area (called the p-n junction or depletion zone). By applying a voltage, the electrons in the n-region gain enough energy to overcome the depletion zone. As soon as these electrons arrive at the p-region, they unite with the positive charges. Energy is released, which is discharged in the form of electromagnetic radiation. The semiconductor material determines the color of the light emitted. The colors red, green, yellow, and blue can generated. The materials are produced on the basis of aluminum indium-gallium phosphide or aluminum gallium arsenide (AlnGaP or AlGaAs) for red and yellow LEDs, and (indium) gallium nitride (InGaN or GaN) for green and blue diodes (FGL 2003).

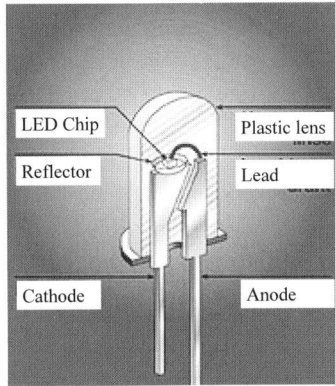

Fig. 42. Structure of an LED[88]

There are two ways to produce white LEDs: By the additive mixture of colors or by luminescence conversion. In the first case, a white LED is produced by combining chips emitting light of different colors into one LED. The various emitters (e.g. red, blue, and green) are placed together so tightly in the device that from a sufficient distance they cannot be distinguished by the human eye and appear as one color; thus producing the impression of white light. This method is also suitable for producing many other colors of light. LEDs producing colored light in this manner (including white) are also called multi-LEDs. The white light produced by multi-LEDs, however, is not as convincing as that produced by means of luminescence conversion. The rendition of the color quality using the multi-LED is poorer; the differing brightness and operating conditions of the various LED chips make this a complicated and therefore expensive solution. Generation of white light by luminescence conversion is achieved us-

---

[88] Source: FGL (2003)

ing a combination of a blue or ultraviolet LED and a luminescence dye. When current is flowing, part of the short-wave light is absorbed by the dye, exciting it to emit light. Yellow-orange (long-wave, low energy) light is emitted. The superimposition of various spectral colors is perceived as white light.

Only with the development of white light LEDs, did the devices first become interesting for illumination purposes. The first white light LED was developed by the Japanese firm Nichia in 1995 and was based on the blue LED, also developed by them; it has been commercially manufactured since 1997. At the same time as Nichia, the Fraunhofer Institute for Applied Solid-State Physics (IAF) developed a white light LED as well as – in close cooperation with Osram OS – a manufacturing process. The know-how was transferred there and production began in summer 1998. Agilent (Lumileds) likewise began mass production of white LEDs in summer 1998. General Electric (GELcore) and Toyoda Gosei have been in production since 1999. White LEDs are now offered by almost all major manufacturers. Today LEDs are increasingly used for the most varied lighting purposes. Table 34 lists possible applications of LEDs according to their properties.

Table 34. Properties and applications of light-emitting diodes[89]

Properties	Application
UV- and IR-free illumination	Museum and showcase lighting
	illumination of fine art
	refrigerated display cases
	medical lighting
	illumination of light-sensitive materials
Low-voltage	damp and subaqueous locations
	portable applications
Small-size	furniture lighting
	directional lighting
	displays
	surface-mount applications
	area lighting, cabinet lighting
Low-temperature	furniture lighting
	refrigerated display
	casesexplosion-hazard environments
Energy saving and long service lifestationary or mobile	transit shelter lighting
	emergency and exit lighting
	garden lighting
Durability	mobile applications

[89] Source: Haller (2003)

4.5 Case study 4: Nano-applications in the lighting industry   143

mobile devices with task illumination
public spaces
signal and emergency lights
maintenance (repair) lighting

4.5.2.3 Scope of investigation and availability of data

The scope (system boundaries) of the comparative assessments includes the entire life cycle of the light sources. The individual life cycle stages include:

- Raw material procurement, pre-production
- Manufacture of light sources
- Use phase
- Disposal/recycling

Data from the most recently available research study was used for the assessment of incandescent and energy-saving lamps. The study, authorized by the Federal Energy Office of Switzerland, was completed by Mani and published in August 1994. Mani worked closely with the manufacturer OSRAM to collect the required data. Specifically, the data on raw materials procurement und light source manufacturing were taken from this study. For the third variant, white LEDs, quantitative data was only available for the production of the 0.35 g semiconductor chip, the core of the LED. The data on LED chip production that included pre-production and raw materials procurement data were taken from the GABI database and are for 1999. No detailed assessment data were available for the other components of the white LED (the housing, series resistor, fluorescent material, etc.). This must be seen in perspective, however, as the greatest share of the environmental impact results from the use phase.

In the submitted comparative assessment of incandescent and energy-savings lamps (Mani 1994), a 75-watt incandescent lamp manufactured by OSRAM (variant 1) was compared to a 15-watt energy-saving Dulux EL lamp with integrated electronic ballast also by OSRAM (variant 2a). As the performance parameters of energy-saving lamps have regularly been improved (particularly with regard to service life), the case study also considers the current energy-saving lamp Dulux EL Longlife (variant 2b), however the production data is based on the former Dulux EL. These conventional lamps are compared to two white LEDs, the "white LED of today" (variant 3a) and the "white LED of tomorrow" (version 3b), each having a power rating of 1 watt. These two variants make possible an assessment of currently available LEDs (assumed luminous efficiency: 18 lm/W) as well as the potential of future efficiency enhancements (assumption: 65 lm/W).

A defined quantity of light was established for the assessment profile. The average quantity of light emitted by the Dulux EL energy-saving lamp from Mani, 6,579 million lumen-hours (lmh), was established as the reference value and functional unit for the profile. The result is arrived at by multiplying the service life of the Dulux EL by the mean luminous flux. The resulting technical data for the five variants used in the case study can be found in Table 35.

Table 35. Technical data utilized for the case-study variants[90]

	Var. 1 incandescent lamp	Var. 2a Dulux EL 15	Var. 2b Dulux EL Longlife	Var. 3a white LED of today	Var. 3b white LED of tomorrow
Power consumption (W)	75	15	15	1	1
Service life (h)	1.000	8.000	13.000	15.000	15.000
Luminous flux (lm) specified	960	900			
Luminous flux (lm) average	897.5	822.4	900	18	65
Luminous efficiency (lm/w)	12	55	60	18	65
Quantity of light per life cycle (Mlmh)	0.898	6.579	11.7	0.27	0.98
Reference quantity (Mlmh)	6.579	6.579	6.579	6.579	6.579
Quantity of lamps used for reference quantity	7.33	1.00	0.56	24.36	6.71
Weight of light source including packaging (g)	42.8	177.1	177.1	0.35*	0.35*

* Here only the LED chip is being referenced, not the complete LED light source.

4.5.2.4 Description of the life cycle stages

Raw material procurement and production

A great number of raw materials are used in the production of light sources. In the following, the essential production steps for three of the light sources are shown, with special reference to the production of LEDs. The manufacture of incandescent lamps consists of four major production steps: Production of the glass envelope, production of the tungsten filament and its support structure, production of several small components, and final assembly of all components.

---

[90] Source: Mani (1994), Gabi 4 database (2001) and own calculations

4.5 Case study 4: Nano-applications in the lighting industry   145

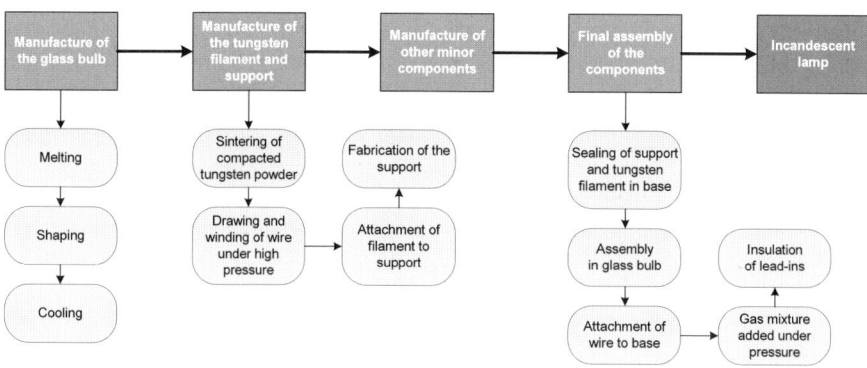

Fig. 43. Stages in the production of the incandescent lamp

The production of compact fluorescent lamps also consists of four major steps: Production of the glass envelope, production of the luminescent material, production of the electrode structure and various small components, and final assembly of the components.

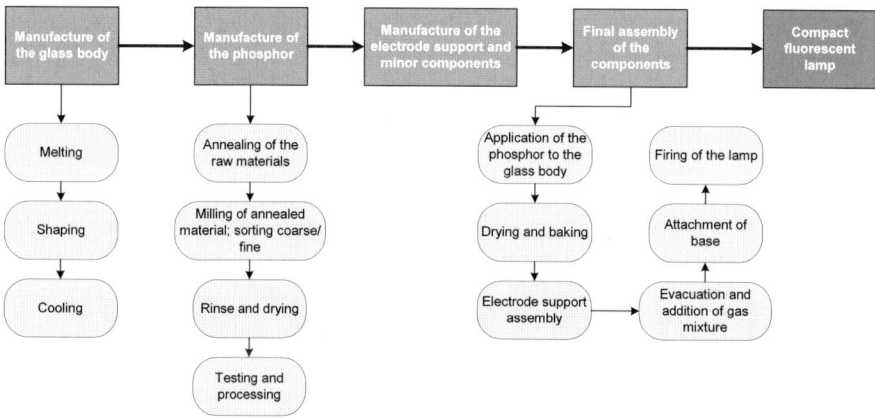

Fig. 44. Stages in production of the compact fluorescent lamp

The production of light-emitting diodes generally follows a different procedure than that described above and is depicted in the following chart. The wafer-processing and subsequent steps will be explained below.

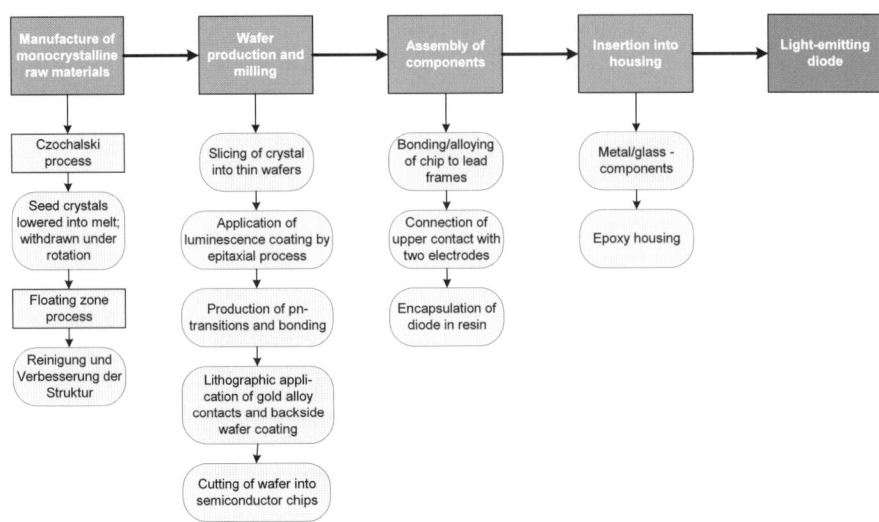

Fig. 45. Stages of production for the LED

The core of a light-emitting diode is the LED chip (consisting of elements from the third and fifth groups of the periodic table), which is produced in the wafer-processing stage. After the mono-crystalline base material is cut into individual wafer-like discs, various semiconductor layers are applied to the wafers using the epitaxial process. This refers to the controlled epitaxial growth of a substance on a mono-crystalline base, the substrate. A key technology in this process, Metal-Organic Chemical Vapor Deposition (MOCVD) has proven to be the best method for the manufacture of optoelectronic and electronic semiconductor layers in modern light sources such as LEDs and lasers, high-output solar cells, and high-frequency electronics and power electronics. Aixtron AG, who builds MOCVD plants worldwide, is the market leader in this field.

In Metal-Organic Chemical Vapor Deposition, nanolayers are grown from gaseous, metallo-organic substances on wafer slices. Pre-reactions of the elements to be deposited are avoided by either conveying them into the reactor in a hydrated form or by means of associated organic molecules (metal-organics). These substances are also called "precursors." The source substances are transported by means of a carrier gas (hydrogen, nitrogen) across a substrate surface in the reactor heated to 350° to 1200°, depending on the material system. In this procedure, the precursor molecules dissociate thermally (pyrolysis) on the substrate surface (Dadgar 2003).

The core of an MOCVD plant is the Planetary Reactor. Geometrically arranged wafers revolve like the planets on a likewise rotating carrier disk.

In this way the wafers are evenly exposed to the process gases. Growth of the layers on the wafer is determined by the quantity and composition of the process gases, temperature, pressure, and time. Up to 50 different layers are needed to produce functional chips; these layers are applied to the wafer during a 5–6 hour production "run." At this point the MOCVD process is complete and the coated wafers leave the reactor. The crystalline deposit of the materials can be very precisely controlled, thus allowing the controlled deposition of the various materials one above the other to create the desired layer properties. A modern blue LED, for example, consists of a great number of different materials such as gallium nitride, used as the base material; indium gallium nitride, for the luminescent layer; and aluminum gallium nitride to enhance efficiency. Silicon and magnesium are used for the n- and p-regions (Krost & Dadgar 2002). In the MOCVD process, by manipulating gas valve switching times and the gas flow, ultrathin layers and abrupt boundary layers can furthermore be produced. The process parameters are easily scaled up to large reactors for mass production. The process can be universally implemented, as there exists a metallo-organic precursor for each chemical element (Grahn 2003).

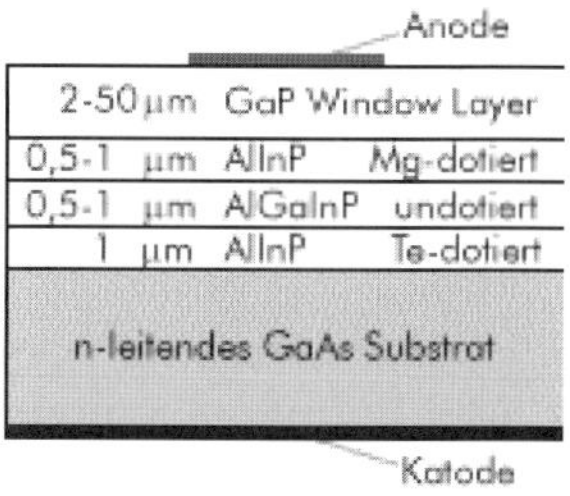

Fig. 46. Schematic representation of the chip structure of an AlInGaP LED

After the wafers leave the reactor, they are characterized (contacts of the pn-transitions). Thereafter electrical contact points consisting of a gold alloy are applied by means of a lithographic photoresist process and an additional layer of metal is applied to the backside of the wafer. To avoid energy losses in the gold alloy, it undergoes an annealing process that lowers its resistance to a fraction of the original value. In the next steps the wafer is attached to an adhesive film, enclosed in a tension ring, and cut with a diamond saw into individual segments or dies 0.35mm by 0.35mm in size. The functionality of the dies is then checked (Aixtron AG 1999). In the case of blue- or ultraviolet-emitting diodes used in the manufacture of white LEDs, the diode chip in the reflector is covered with a drop of luminescent dye. Thereafter the various components of the light-emitting diode are assembled.

LEDs come in a great number of structural configurations. Diverse metal/glass – or more commonly, epoxy or plastic – housings are used. The latter can easily be produced in suitable numbers for mass production.
T-type LED: This form is the original structural shape of the light-emitting diode. It consists of the LED chip, the contacts, gold or aluminum connecting wire, and the plastic housing. After fixing the chip to the reflector on the metal lead, the contact to the second lead is established using a thin gold wire. Finally this is all molded into a stable unit with epoxy resin or other plastic material. The optical characteristics of the LED are determined by the reflector geometry, the shape of the plastic housing, and the position of the chip inside the housing (Vossloh Schwabe Deutschland GmbH 2003).

SMD LED: There are also the LED soldered to the back of the printed circuit board and the SMD (Surface Mounted Device), an extremely miniaturized form of the LED. Unlike the standard LED, the SMD LED has no metallic reflector. There are many different SMD structural shapes and sizes, varying according to application.

COB LED: Tightly packed, effective heat-dissipating solutions can be achieved using chip-on-board (COB) technology. The raw chips are directly installed on the circuit board. The chip connections and the board are connected with gold wire. The chip is covered and protected with a drop of epoxy resin. This technology allows more light to be radiated from a smaller surface.

LED modules are standardized or customized LEDs that consist of several LED-bearing circuit boards. These are readily used by manufacturers of display and signal light technology. The boards to which the diodes are attached can have many possible shapes. By varying the dimensions, mounted components, and wiring, the housing shapes can be varied with relative freedom (Haller 2003). Furthermore, the first LED modules which could someday replace conventional lamps already exist: The socket-mount LED module. It combines LED, necessary electronics, and the standard base (e.g. E27) to form a complete lamp that may be used in an existing luminaire housing.

Use phase

Incandescent and energy-saving lamps are used everywhere: in business and industry as well as in the home; however the incandescent lamp is used much more commonly in the private household. More common in the business place is the energy-efficient fluorescent tube. Because of economic reasons already discussed, white LEDs are still used mostly in spe-

4.5 Case study 4: Nano-applications in the lighting industry

cial applications. In order to nonetheless assess and compare potential environmental impacts for the entire field, white LEDs are compared in this case study to conventional light sources.

The use phase is of major consequence in the life cycle assessment profile, as light sources are a significant consumer of electrical energy in this phase. The energy consumption is based upon the production of a specific reference light quantity (RLQ), which serves as an equivalence value. Specific energy consumptions of the variants are derived from multiplying power input, service life, and number of lamps required.

Table 36. Energy consumption of each case-study variant[91]

RLQ = 6.579 Mlmh	Variant 1 incandescent lamp	Variant 2a Dulux EL 15	Variant 2b Dulux EL Longlife	Variant 3a white LED of today	Variant 3b white LED of tomorrow
Power input (W)	75	15	15	1	1
Service life (h)	1000	8000	13000	15000	15000
Lamps required for RLQ (quantity)	7,33	1,00	0,56	24,36	6,71
Energy consumption for RLQ (kWh)	550	120	109	365	101

Emission factors for the current electrical power mix in Germany as taken from the GEMIS 4.1 database and listed in the appendix (Table 56) form the basis for the calculations of environmental impacts in the use phase (GEMIS 4.1 2003).

Disposal

There are no legal regulations governing the disposal of incandescent lamps, as the environmental impact of almost all substances they contain is negligible due to their nature and the amounts involved. Therefore almost all incandescent lamps are disposed of in the household waste. The disposal of compact fluorescents, however, is regulated. In 1996, the German Waste Avoidance, Recycling and Disposal Act came into effect. It places discharge-type lamps (in the case of disposal and recycling) into the category of waste requiring special attention. The lamps must be collected separately and classified as hazardous waste or be recycled.

---

[91] Source: Mani (1994), Gabi 4 database (2001) and authors' own calculations

The regulations governing the disposal of electrical and electronic devices changed in 2002. Following modifications to the European Waste Catalogue, in January 2002, almost all electronic devices and electrical devices are now classified as waste requiring special handling. They may not be disposed of in household or ordinary waste. The fluorescent tube and the light-emitting diode are also listed in this catalogue. They are classified as materials for special handling and supervision – the energy-saving lamp because of its mercury content and the LED because of possible quantities of arsenic, gallium, phosphor, or compounds containing them (Senatsverwaltung für Stadtentwicklung Berlin 2003). Newer developments include the EU regulations on to-be-disposed-of electrical and electronic devices, in effect since February 2003. The directive 2005/96/EG regulates the disposal of electronic and electrical devices, whereas directive 2005/95/EG limits the use of certain hazardous substances in these devices. The European directives stipulate that the listed products (including fluorescent tubes and LEDs) must be returned to and accepted by the manufacturers and properly disposed of. Additionally, directive 2002/95/EG contains several exceptions with regard to the avoidance of certain hazardous substances in products. For example, it is explicitly mentioned that no more than 5 mg mercury will be allowed in compact fluorescent lamps in the future. The directives are to be incorporated into national law by August 2004; draft regulations for the disposal of electrical and electronic devices, based upon the Waste Avoidance, Recycling and Disposal Act, already exist.

Detailed data for this life cycle stage are included in the environmental assessment by Mani (1994) for the incandescent lamp and the energy-saving lamp. No quantitative data is available for LEDs. As it is clear from the assessment data for the incandescent lamp and the energy-saving lamp that this stage only plays a minor role in the overall investigation, the question of disposal will not considered quantitatively, but only discussed with respect to problematic substances (mercury).

4.5.3 Life cycle inventory analysis

In the life cycle inventory analysis, the material and energy relationships between the lighting system being studied and the environment are recorded, i.e. the input flows from the environment and the output flows that are returned to the environment are noted. The goal is to establish a data inventory based upon functional equivalents for the selected variants. Since quantitative data for disposal are not available, calculations can only be made for manufacturing, inclusive of raw materials procurement, and the use phase.

The total absolute material and primary energy amounts needed for the manufacture of the quantity of lamps required to generate the set reference light quantity (RLQ) of 6.579 million lumen hours are listed in the table below. The total material flows for the five lighting variants can be found in Table 57 in the appendix.

Table 37. Material and energy requirements for the production of the lamps for the RLQ[92]

RLQ = 6.579 Mlmh	Variant 1 incandescent lamp	Variant 2a Dulux EL 15	Variant 2b Dulux EL Longlife	Variant 3a White LED of today	Variant 3b White LED of tomorrow
Material requirements	669.4 g	453.3 g	253.8 g	2671.5 g	735.9 g
Product quantity, total	313.7 g	177.1 g	99,2 g	8.5 g*	2.3 g*
Primary energy requirement	14.7 MJ	32.6 MJ	18.3 MJ	5.9 MJ	1.6 MJ

* Here only the LED chip is being referenced, not the complete LED light source.

In looking at the material quantities for the first three variants, consideration must be given to the packaging. The ratio of materials required to product quantity is 2.1–2.6:1. The large material quantities for LED chip production in proportion to product quantity reflect the fact that semiconductor technologies require considerable volumes of raw materials such as ore and stone and auxiliary materials in order to produce a small, highly complex quantity of product. The ratio is 314:1. Moreover, an assessment of a complete LED lighting system must also consider materials for the housing, ballast resistor, packaging, etc.

The situation is different for the primary energy requirement. The primary energy requirement for LED chip production is lower than that for conventional light sources. The difference must again be put into perspective, as the missing system components in the LED lighting system must be included. Production of the circuit board for the ballast for the Dulux EL15 energy-saving lamp alone consumes 17.3 MJ of the indicated 32.6 MJ/BLM. It can therefore be assumed that the primary energy requirement of an LED lighting system is comparable to that of conventional lighting sources.

---

[92] Source: Mani (1994), Gabi 4 database (2001) and own calculations

Table 38. Energy requirements for the use phase for generation of the RLQ[93]

RLQ = 6.579 Mlmh	Variant 1 incandescent lamp	Variant 2a Dulux EL 15	Variant 2b Dulux EL Longlife	Variant 3a white LED of today	Variant 3b white LED of tomorrow
Primary energy requirement	6.271 MJ	1.368 MJ	1.243 MJ	4.161 MJ	1.152 MJ

If energy consumption in the use phase is compared to that of the production phase, it becomes obvious that, depending on the variant, 97–99% of the energy is consumed during the use phase, with the result that the deviation caused by the incomplete consideration of the LED lighting system can be viewed as minimal. The central measure for the environmental assessment of light sources being used for illumination is energy consumption during the use phase and the associated emissions. Energy consumption for raw materials procurement und manufacture of the light sources is minimal. Moreover it becomes very clear that the current white LED is at a disadvantage by a factor of 3 as compared to the energy-saving lamp. Only, if the future scenario for the white LED comes to pass, i.e. a luminous efficiency above roughly 65 lm/W is achieved, will energy consumption become comparable to energy-saving lamps.

This fact is also generally confirmed by the calculated emissions, which accumulate as relative quantities. Here again, emissions resulting from power consumption during the use phase dominate. These emission quantities are shown in brief in the table below and in detail in Table 58 in the appendix.

---

[93] Source: Mani (1994), Gabi 4 database (2001) and author's own calculations

4.5 Case study 4: Nano-applications in the lighting industry 153

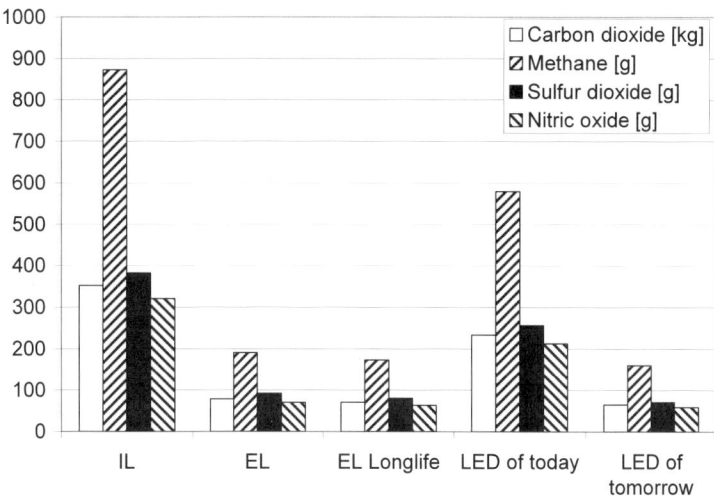

Fig. 47. Selected emission quantities of the case-study variants relative to the RLQ[94]

The total emission quantities expressed in the bar chart also make clear that the white LED of today is better than the conventional incandescent, but at a disadvantage by a factor of three when compared to the energy-saving lamp. Only if the future scenario for the white LED is realized, will the emissions become comparable to those of energy-saving lamps.

With respect to critical substances used in the light sources, mercury (found in the energy-saving lamp) and arsenic (used in the manufacturing process for the white LED), in particular, must be considered. Technological improvements in recent years have made it possible to significantly reduce the proportion of mercury contained in fluorescent tubes. The energy-saving lamp Dulux EL by OSRAM contained as much as 10 mg of mercury in 1994; currently the compact fluorescent by OSRAM contains roughly 4 mg of mercury; this represents its emission potential, in the case of release caused by improper disposal. If lamps break at the waste disposal site, mercury can escape directly into the environment. Mercury and numerous mercury compounds are volatile and highly poisonous, which is one of the main reasons for disposing of these lamps as hazardous waste or recycling them. Furthermore, mercury emissions in the production phase and the use phase must be considered in the overall assessment, as emissions also take place, for example, during power generation.

---

[94] Source: authors (database: authors, Mani 1994, Gabi 4 database 2001)

Table 39. (Potential) mercury emissions of the case-study variants[95]

RLQ = 6.579 Mlmh	Variant 1 incandescent lamp	Variant 2a Dulux EL 15	Variant 2b Dulux EL Longlife	Variant 3a white LED of today	Variant 3b white LED of tomorrow
Hg emissions in raw materials procurement / manufacturing (mg)	0.004	0.26	0.15	0.03	0.01
Hg emissions in use phase (mg)	6.65	1.45	1.32	4.41	1.22
Hg quantity contained in product and potential release risk (mg)		10.00	2.24		
Total (mg)	6.65	11.71	3.70	4.44	1.23

The (potential) mercury emissions mostly arise during the use phase. Only in energy-saving lamps are considerable emission quantities added by the mercury contained in the product itself. Here it is clear that current energy-saving lamps with respect to total quantity still are better than current white = LEDs. Only in the white LED of tomorrow scenario would a significant avoidance of mercury emissions be possible.

No quantitative data for arsenic and its compounds was available for evaluation. The most common materials for the production of LED chips are aluminum indium gallium phosphide and aluminum gallium arsenide (AlInGaP and AlGaAs) for red and yellow LEDs, and indium gallium nitride (InGaN and GaN) for green and blue diodes (FGL 2003 and others). Since the production of white LEDs requires either blue, green, and red diodes or else blue-emitting diodes, these substances might also be present in white LEDs. Gallium arsenide, like gallium nitride and gallium phosphide, belongs to the semiconductor groups III/V. With respect to handling and disposal, arsenic and its compounds are of the greatest significance. The toxicity of arsenic and its compounds varies greatly, but the substances used in the semiconductor industry tend to represent a certain hazard potential. The carcinogenicity, mutagenicity, and reproductive toxicity of many arsenic compounds is indisputable (BLU 2002).

As a semiconductor material, gallium arsenide is not poisonous. But in the presence of oxygen and water, an ultra-thin, very toxic layer may form

---

[95] Source: Mani (1994), Gabi 4 database (2001) and authors' own calculations

on the surface of the material, which could cause environmental damage at a conventional landfill disposal site. Furthermore, an extremely poisonous gas is produced in the manufacture of gallium arsenide: arsine (arsenic hydrogen), chemical formula AsH3, which is used to guarantee the purity of the semiconductor material. Even minimal concentrations of a few arsine molecules per million gas particles in the air can cause severe health damage or be even lethal. The arsenic-hydrogen bond is highly toxic. It blocks nerve receptors and can impede the transport of oxygen in the body.

Therefore, light-emitting diodes also require special disposal treatment and supervision due to the possibility that they may contain arsenic, gallium, and phosphor compounds. According to the European directives, listed products must be accepted in return by the manufacturers and properly disposed of. A recycling procedure for LEDs does not yet exist (Grezcmiel 2001). Researchers at the University of Marburg found an alternative to arsine years ago. The alternative substance is less volatile and has a much lower vapor pressure. It is considerably less harmful for the environment than arsine. The substance is of interest for semiconductor industry, because the much lower hazard risk reduces the costs of a semiconductor production facility. Waste quantities are also much lower (Thimm 1999).

4.5.4 Life cycle impact assessment

To complete the impact assessment, it is necessary to have access to emissions data which can be allocated to specific environmental impacts. Since the data shows a direct relationship between energy consumption and the relevant emissions, this would also be seen in the categories greenhouse effect, acidification, and eutrophication and not reveal any new knowledge; we therefore refrain from making this presentation.

4.5.5 Light sources based on quantum dot technology

4.5.5.1 Definition and background physics

Quantum physics describes particles having wave functions. This particularly applies to electrons, as the smallest stable particles with a rest mass. Should the wave function of the electron be increased until it reaches the geometric material realm, quantum mechanical effects can be anticipated, leading to interesting new properties.

This holds true for nanoparticles below about 20 nm in size (typically 1–5 nm). They are at the borderline between individual molecule and dimensional crystal. The movement of the electrons is constrained by the minuteness of the nanoparticles. Since the electrical and optical properties of solid bodies are determined by their electrons, a discretization of energetic states and a broadened band gap between valence band (completely filled with electrons) and conduction band (not completely filled so electrons are mobile) can be observed with decreasing material size (Haase & Kömpe 2003). The resulting new properties can only be explained with the help of quantum physics. Such nanoparticles are therefore called quantum dots and are more or less zero-dimensional.

The electrons feel "squished" by the restricted particle boundaries and thus the distance (band gap) between original state and excited state increases with the decreasing size of the particles. If an electron shifts from the excited state into the original state, light is emitted; its wavelength and energy level are dependent on the band gap. It is therefore possible to generate wave lengths ranging from UV through the visible spectrum on up to the infrared (350–2300 nm) using quantum dots of the same composition (Evident Technologies 2003). The optical properties are determined by the size of the quantum dots. This is what makes the quantum dot so interesting for lighting technology. Earlier it was only possible to generate light of different wavelengths by combining different substances. Quantum dots even permit using substances for light generation that were previously unsuitable (Bertram & Weller 2002). Additionally, more than 50% of the atoms lie at the surface due to the small size of the quantum dots, thus permitting a precise tuning of light-emitting properties and suggesting that the emission of several colors from a single dot may be possible (SNL 2003). Quantum dot crystals used in combination with other crystals or phosphors can emit any desired color.

Quantum dots can be produced technically by means of epitaxial growth, for example in vacuum precipitation processes or by chemical methods from colloidal solutions (Rubahn 2002).

4.5.5.2 Applications of quantum dots

By irradiating a solid body with light of a suitable wavelength, electrons can be advanced from the valence band to the conduction band. The energy input must be at least as great as the band gap. After a short time, the electron falls back to the valence band, with the surplus energy often being emitted in the form of light. The energy or the color of this emitted fluorescent light corresponds to the band gap energy. Several methods are being applied for using quantum dots to produce light.

Direct-charge injection: In direct-charge injection, electrons in the quantum dots are brought into the excited state by the transport of electrons or electron voids and the resulting collision processes. Research on this method is taking place at the Massachusetts Institute of Technology (MIT, Cambridge) and elsewhere; in 2003 they succeeded in making a major step forward in development. The problem in direct-charge injection is to stimulate as many of the electrons flowing through the carrier material as possible to electron excitation in the quantum dots and thus light production. MIT researchers were able to augment the efficiency of direct-charge injection by a factor of 25 by placing CdSe dots (cadmium selenide) between two organic layers (Riebeek 2003). With this technology an efficiency increase of up to 100% (all injected charges generate light) may be possible, more than with any other light source. This would raise the energy efficiency of light sources to a new dimension. The colors being emitted are two to three times as pure as those of ordinary OLEDs (Riebeek 2003). After further development, this technology, an extension of OLED display technology, will revolutionize flat-screen technology. The durability and stability of quantum dots are positive factors in this respect.

Stimulation by UV light: Researchers at the Department of Energy's (DOE) Sandia National Laboratories, Albuquerque, have chosen a different approach. They developed the first white light-emitting device using quantum dots (SNL 2003). In contrast to direct-charge injection, stimulation is effected by UV light (380–420nm wavelength) rather than a current. The quantum dots are covered and encapsulated with suitable organic molecules such that they emit light in the visible spectrum when stimulated by ultraviolet light. The required UV light is produced by traditional LED technology. Therefore the efficiency cannot be higher than that of blue-light LEDs. However, it is higher than in traditional white-light LEDs, as their phosphor materials show poor absorption for the blue wavelength and thus low efficiency. Optical backscatter losses of traditional phosphor materials reduce the efficiency by 50%. This can be overcome with the new technology. In 2004, researchers tried to increase the concentration of quantum dots in the capsules to enhance light efficiency and to understand the behavior of quantum dots in higher concentrations.

So far, researchers at MIT and Sandia have mainly used quantum dots of the semiconductor materials cadmium sulfide and cadmium selenide, which contain the poisonous heavy metal cadmium and are therefore not appropriate for mass production. Harmless alternatives, however, are being investigated, including nanocrystalline silicon and germanium with a surface of light-emitting manganese ions. The Nanoelectronics Research Centre (NRC), at the University of Glasgow, has already produced Si-SiGe quantum-dot-based LEDs, which according to the researchers, would be ideal as light sources in optical circuits, because they can be directly integrated into the Si chips (Tang et al. 2003).

In addition to the direct generation of light, quantum dots are also used as markers in fluorescence microscopy of biological and medical preparations. They are tied to biomolecules (e.g. DNA fragments) and stimulated to emit light of various colors when irradiated with UV or blue light. Compared to current dyes, quantum dots are more stable and tend less to fade. This allows more detailed investigation of a preparation and the differentiation of details becomes easier. By using quantum dots, different parts of the preparation can be colored differently, thus allowing the recognition and determination of various parts of the tissue simultaneously (Bertram & Weller 2002). Use of quantum dots promises faster, more flexible, and less expensive tests and immediate biological analyses and patient diagnoses. However they must be prepared with sufficient precision if they are to be used in medicine and research.

Components containing quantum dots also promise improvements in the development of lasers, detectors, optoelectronic switches, and memory elements (e.g. optical high-density memories). Nano-electronics also has great hopes for the application of quantum dots. Due to their high quantization energy, nanoparticles are capable of storing discrete amounts of electrons (charge carriers) (Bertram & Weller 2002). Attempts to use quantum dots as memory devices are already being made by several firms. According to Bertram and Weller (2002), a combination of quantum dots as memory cells with a molecular switch leading to a braid of nanowires would be the first step on the way to the nanocomputer. Generally it can be stated that research work around the world on quantum dots technology has recently greatly intensified.

4.5.5.3 Summary and prospects for the implementation of quantum dots

Quantum dots are nanoscale particles at the boundary between molecule and solid body. Their composition and tiny size are responsible for their extraordinary optical properties, which may be adjusted to meet specific requirements by changing the size and chemical composition (of the surface). Under stimulation they can be induced to emit light. In contrast to other light sources, they generate an extremely pure and bright light and can at the same time emit an entire spectrum of colors by stimulation of a single wavelength.

It is anticipated that quantum dot technology will have a firm place in display technology in the long term, especially in combination with OLEDs, however their development is not yet as far along as that of the

OLED. The use of quantum dots will further enhance the efficiency of light sources, generate more brilliant colors, and reduce the number of necessary manufacturing steps in display manufacture. These are economic and environmentally relevant advantages in favor of quantum dots and the use of their optical properties in lighting technology. However, an industrial application is not to be expected in the near future.

4.5.6 Case-study Summary

The objective of this case study was an investigation of the eco-efficiency potential of new nanotechnological products in the lighting industry. For this purpose, white LEDs were compared to conventional light sources including the incandescent lamp and the energy-saving lamp (compact fluorescent), and the future potential of quantum dots was considered.

The case study demonstrates very clearly that energy consumption during the use phase is by far the most significant factor in the environmental assessment of light sources for lighting purposes. The current white LED compares more favorably to the classic incandescent lamp, but is at a disadvantage by a factor of three when compared to the energy-saving lamp. Only at luminous efficiency values above approx. 65 lm/W, values which were assumed for a future white LED, will they be able to compete with the energy-saving lamp with respect to environmental impact. This was confirmed by Arpad Bergh, president of the US Optoelectronics Industry Development Association (Interview by Siemens NewsDesk 2003). He makes the assumption that the luminous efficiency of the white LED must be increased to 85–100 lm/W before it will become interesting for everyday illumination purposes.

It can be assumed that the use of quantum dots in the future will make possible increases in the energy efficiency of light sources. Quantum dot technology is expected to have a firm place in display technology over the long term, particularly in combination with OLEDs. Actual application of quantum dots in commercial products, however, is still some years away.

4.6 Case study 5: The risk potential of nano-scale structures

The following section presents and discusses the potential risks of nanotechnological applications. The discussion specifically focuses on nanoparticles, as nanoparticles are already being produced and applied on a large scale and thus form the basis for numerous other applications; furthermore, relatively broad discussions on the risk potential of nanoparticles are already taking place. The chapter consists of four parts.

The first part looks at those risks that already can be expected on the basis of the properties of intentionally manufactured nano-scale products, agents, and materials; furthermore, a literature survey on the problematic effects of selected substances and structures for humans and the environment is presented. The second part, taking as its example the nanoscalar titanium oxide found in suntan lotions, discusses the problem of equating substances of micro- and nano-size. The third part analyses the life cycle of nanoparticles. Finally, we compile the results of the study and illuminate possible consequences of these findings. In the course of the case study, a written survey was conducted and experts in the field of nanoparticle toxicology from Great Britain, Germany, and the United States were polled. The results are noted in the relevant sections of this chapter. A detailed presentation of the survey results can be found in the appendix.

4.6.1 Potential risks of nanotechnology

Development of the potential of nanotechnology is only at its starting point, with relatively simple applications and products already entering or at the threshold of the market, many of which involve the application of nanoparticles. One example is nanoscalar carbon black, added to automobile tires to improve abrasion resistance, which has already been in use for a very long time. In many cases, the examples discussed can also be viewed as a continuation of lines of decades-long development. Many of these examples still compete with older traditional solutions. In such cases, nano-scale innovations are thus only replacing or enhancing already existing conventional solutions. The novelty of these examples is for the most part only to be found in the nanoscalar size of the particles.

Nonetheless, the question arises as to the extent to which the drive into the area of nanotechnology entails new types of effects. Do nanoscalar materials have new (or do they enhance already known) properties that could be detrimental to the environment and/ or human health? Such new and/or enhanced effects are normally to be expected. After all, it is such altered or enhanced properties that make nanomaterials interesting for production purposes.

However, questions concerning new (or enhanced) properties or effects and possible negative side effects and consequences have so far rarely been systematically investigated or formulated. The neglect of side effects is partly the result of institutional delay. Established procedures for regulating hazardous substances have not yet been adapted to address these new issues. The non-governmental organization ETC Group (ETC Group 2002), for example, notes that the addition of considerable quantities of titanium dioxide nanoparticles to high-sun-protection-factor suntan lotions required no new investigations in the USA. Conventional titanium dioxide had already been tested and approved; new applications of titanium dioxide nanoparticles were considered to be equivalent to pre-nanotechnology applications with respect to environmental and health impacts (see ETC Group 2002). Whether this equivalency is justified, cannot be settled here, but in light of the altered properties found at the nano-level, it should at least be addressed.

This is of particular importance, knowing as little as we do, so far, about the environmental and health aspects of nanoparticles. Those investigations that have been conducted on the effects of ultra-fine particles resulting from combustion processes are alarming.

However it should be noted that the common properties of nano-scale systems are predominantly due to the scale of the materials and products. This means, on the one hand, that in nanotechnology's role as a cross-sectional technology, problems may occur that are primarily "contingent on the technology." On the other hand, other problems may arise from a specific combination of circumstances within the context of specific applications. The technologies as well as the application contexts may greatly diverge and likewise the potential risks to environment and health that need to be addressed. For example, nanotechnological procedures and products and their impact in the field of biotechnology will differ fundamentally from those in areas such as electronics and thin-film technology. Moreover, different impacts on the various environmental compartments can also be expected.

4.6.1.1 Technology-specific characteristics of nanoparticles as possible hazard sources

An initial preliminary technical characterization of nanoparticles and nano-structured surfaces reveals the following lowest common denominators. It can be stated, that:
- We are dealing with structures in the nano-dimensional realm
- Properties of the molecules are altered in this dimension (thus leading to desired effects and behavior)
- In particular, the ratio of surface to volume changes

Additionally, one must specifically mention that substances such as fullerenes and nanotubes do not generally exist in the natural environment. The impact of such new materials on environment and health is difficult to foresee.

In addition to the properties of nanoscalar materials already mentioned, the following aspects are of particular importance:
- Mobility in environmental compartments and inside the organism
- Reactivity and reactive specificity
- Bioaccumulation
- Persistency
- Non-occurrence in nature
- Pulmonary intrusion
- Water solubility, liposolubility
- Carrier and piggyback effects
- Agglomeration, dispersion
- Other properties

As science learns more about conditions of implementation and application contexts, other aspects must also be considered, for example:
- Nature of application and quantities
- Contained / non-contained (open/closed) applications
- Aspects of the life cycle such as raw material expenditures, recyclability…

In an attempt to characterize nanomaterials in a standardized manner, we can make the following statements: Nanotechnological systems and products, particularly nanoparticles, for the most part currently represent the further development of known processes and products.

The key difference is that it is increasingly possible to shape things at the atomic and molecular level. This means several scientific and technical disciplines can contribute to the production of nano-scale building blocks and components.

Industrially produced nanomaterials cannot be viewed as a uniform substance or materials group, thus making the assessment of potential risks and hazards even more difficult. There is, in fact, an enormous range of structures and materials in the nanoscalar dimension. Their distinguishing characteristics:
- A large number of chemical elements and compounds are utilized
- They exist at the nano-scale level in various sizes and have various surface structures
- As yet, no particularly worrisome group of substances to be singled out for investigation has been identified.

Consequently there can be no simple answer to the question, as to whether nanoparticles are safe and what impact they will have on human life and the environment. The differences in size, shape, surface, chemical composition, and biopersistency require that each nanomaterial be individually investigated with respect to possible environmental and health hazards. Very similar compounds may have very different effects. It is also known from toxicity research that certain substances may be comparatively harmless as long as they are applied to the skin or taken orally, but can be extremely toxic when inhaled (Colvin 2003a, Hoet 2004). The question, as to whether and to what extent general statements on hazards caused by specific substance groups or structures can already be made, will be dealt with later in this paper.

Nanoparticles
Nanoparticles can be classified as belonging to the transitional area between the realm of individual atoms and molecules and that of larger ensembles.

Many of the particularly interesting new properties of nanoparticles are due to the altered ratio of surface to volume. Because of their incredibly small size, nanoparticles have an enormous surface relative to their volume. They are, therefore, very reactive, reacting comparatively quickly and aggressively, in part even spontaneously with their environment. The reason for their reactivity is found in the surface of the nanoparticle. Due the altered surface volume ratio, the surface of the nanoparticle has many more free electrons, which, due to their position on the surface, are capable of reacting with their environment. Thus nano-scale materials and substances are more reactive than structures having a smaller relative surface area. For this reason, existing, known substances that now can be produced in nano-sizes can suddenly have properties and effects very different from those of their larger counterparts.

Some experimental results have already given rise to the question as to whether nanoparticles because of their tiny size and pulmonary intrusiveness and low biodegradability alone may provoke certain toxic effects almost independently of their specific composition. Other findings have contradicted this thesis. Asked whether the effects of nanoparticles are rather the result of their size, or their chemical composition, or the nature of the surface, the experts surveyed in the course of this project responded:

- Negative effects are due to size, as well as chemical composition, and surface structure. A weighting of the three aspects is currently not possible.
- Likewise, dosage, coating, shape, distribution in the tissue, distribution of charge or electrical potential between molecules, and degree of agglomeration may also have an influence.
- Effects are also determined by the condition of the human body: condition of the immune system, intake path (skin, respiratory, direct injection, etc.), and general state of health.

- One survey participant held the opinion that there might be differences between the effects of individual free nanoparticles and those bound in, for example, a composite material.

Nanoparticles are also capable of adsorbing molecular contaminants.

Adsorption can assist foreign matter in gaining access to parts of the body and the cell, from which it otherwise would be blocked. Transport of such biological contamination could be a greater risk for biological systems than nanoparticles as such, scientists say (New Scientist 2003).

Behavior of nanoparticles released into the environment

The problem of assessing nanoparticle effects is further complicated by the fact that a possible negative impact also depends on the agglomeration behavior of nanoparticles. It is known from the behavior of ultra-fine particles in combustion processes that they already begin to attach to each other shortly after combustion to form larger groups of particles (agglomerates). Oberdörster points out that agglomerated particles are neither more nor less problematic than other particle forms, but that individual nanoparticles can cause severe problems. It is likewise known that nano-scale titanium dioxide agglomerates more easily in aqueous than in lipophilic solutions. So far, it has not been possible to make generalized assumptions about the behavior of nanoparticles in differing media. This statement is also confirmed by our expert survey. General statements of the kind "All fullerenes agglomerate" cannot be made. The statements given on common behaviors are far less general and were all qualified:

- Nanoparticles can definitely form aggregates in gaseous and liquid media.
- Aggregates spread less evenly than individual nanoparticles.
- Airborne particles agglomerate homogeneously within ten seconds if the concentration is relatively high (10^7–10^8 particles/cm^3). Lower concentrations need a considerably longer period of time.
- Agglomeration behavior depends on the surface coating, chemical reactivity, and electrical potential.
- If through the accretion of further atoms and molecules larger agglomerates are formed, the total surface area and thus the reactivity of the agglomerate may again change.
- The larger the surface, the greater the possibility that substances will aggregate. Agglomeration is also determined by the surface reactivity.
- Aggregated particles may de-aggregate again in the human body.
- Toxicity of particles may deteriorate during their life span.

Ultra-fine particles and nanoparticles
The presently limited knowledge of expected and possible effects and concerns about the negative effects of nanoparticles is based partly on analogies to the knowledge on ultra-fine particles. Ultra-fine particles (PM0.1) are the same size as nanoparticles; however, they are not industrially manufactured, but are the result of combustion processes. Ultra-fine particles have an average diameter of less than 0.1 micrometer (μm).

Epidemiological investigations of larger particles (diameter less than 10 micrometers) suggest with comparatively strong evidence that chronic exposure to particles in the air has a harmful effect on the cardiovascular system (Dockery et al. 1993, Kunzli et al. 2000). It is presently under discussion whether ultra-fine particles automatically cause greater damage than do larger particles because of their greater surface. Only a few studies on ultra-fine particles exist, but there are indications that such particles are more dangerous than larger ones (Howard 2004b).

Indications for negative effects of ultra-fine particles were also found by in-vitro investigations. Diabeté (2002) writes: "By means of in-vitro tests it was shown that ultra-fine synthetic particles have a stronger cytotoxic effect than larger particles of the same chemical composition. Flue dust from a waste incineration plant containing nano-scale particles increased the release of pro-inflammatory cytokines in lipopolysaccharid-stimulated macrophages and inhibited the formation of NO radicals. In co-cultures of macrophages and pulmonary epithelial cells it was shown that they release more cytokines than the total of the respective individual cell cultures." These indications may suggest that nanoparticles, which are in part even smaller, in similar concentrations could a priori likewise be harmful to the cardiovascular system. Since ultra-fine particles are combustion byproducts, equating the effects of the two types of particles can only be done to a limited extent. Nanoparticles may cause additional and/or other problems (Colvin 2003a, Kreyling et al. 2004).

The above statements were confirmed by the results of the survey. According to the experts, substantial empirical data on the toxic impact of ultra-fine particles in the context of air pollution caused by articles of natural and combustion origin exist. The effects of ultra-fine particles on the pulmonary system, the cardiovascular system, and human blood are well documented. Claims about possibly harmful effects of inhaled particles from combustion processes, however, do not allow any direct conclusions about industrially manufactured nanoparticles. An example for a possible similarity between both classes of particles is the capability of ultra-fine particles to bypass the body's defense mechanisms, penetrate cells, navigate through the body, and cause inflammatory reactions. Findings derived from particle toxicology thus point to concerns which should not be disregarded. On the other hand it is necessary to develop new methods of investigation and assessment that are suitable for industrially produced nanoparticles.

4.6.1.2 Impact on health and environment

The following summarizes conclusions that suggest some possible effects. Results from relevant studies and the more or less substantiated conjectures of researchers in this field are also presented.

These statements are meant to be representative only – they could be supplemented by more examples – even so, no general conclusions about nanotechnology can be drawn from them. What does become clear, however, is that this is a field with a high degree of uncertainty and a large number of unknowns, whose systematic assessment has not yet begun.

Effects on the human body

Nanoparticles are very mobile: in animal subjects they have entered the liver, the brain, and even the fetus. Howard (University of Liverpool, England) reports a possible migration of nanoparticles into the fetus. His analysis was not yet available at the time of publication, but he was confident that evidence to support his initial conclusions would soon be available (Howard 2004a). Howard assumes that there is a natural transport path for nanoparticles, which they use to enter the body and move within it. He assumes that nanoparticles pass through the caveoles (openings) of the cell membrane. The openings have a width between 40 and 200 nm and seem to play a role in transport of macromolecules and proteins. Caveoles are big enough to transport nanoparticles (Howard 2004b). There is nothing known about possible effects in the body. One should add that the particles investigated by Howard were specifically prepared to penetrate such membranes.

Oberdörster (University of Rochester, New York) traced the distribution of carbon particles of 35 nm diameter after applying them to the nasal mucous membrane of rats. One day later the nanoparticles were found in the brain. The ends of the olfactory nerves had absorbed the particles and transported them into the bulbus olfactorius. Concentrations increased until the experiment was stopped seven days later. It is unclear what effects the particles have on the brain (Oberdörster 2004). Oberdörster reports in the context of the survey, that the same effect appears with other substances (30 nm virus, 50 nm colloidal gold, 30 nm Mn oxide) and also if applied to primates. Mn oxide caused reactions in rats that suggested inflammation at a concentration of 450 µg/m^3 and twelve days' exposure. In a somewhat older study, Öberdörster et al. reported carbon, between 20 and 29 nm in size, being found in the livers of rats after six hours of inhalation (Oberdörster et al. 2002). The particles entered the blood circulation via the lungs. Oberdörster et al. emphasize that translocation into the blood and the organs may vary with other nanoparticles.

Nanoparticles may enter through the human lungs because they are too small to be filtered or to be destroyed by the protective mechanisms of the respiratory system (pulmonary mucus and macrophages). The pulmonary mucus that lines the lungs like a carpet is unable to filter fine and ultra-fine particles from the air or to transport them upwards. In the alveoli that lie beneath, where the gaseous interchange between lungs and blood occurs, macrophages absorb undesirable substances. Macrophages do not recognize structures/substances smaller than 70 nm as matter extraneous to the body.

The problem of nanoparticle translocation may become aggravated; as Donaldson (Napier University, Edinburgh) commented in the survey, if – as planned – nanoparticles are added to the human blood for medical treatment in the future. However, medical applications would be subjected to much stricter regulations and testing than the particles in industrial production that are being considered here.

There is likewise the suspicion that nanotubes could be harmful for the human organism. Lam introduced 0.1 and 0.5 mg carbon nanotubes into mice via the windpipe. After seven days all test animals had developed granulomae and in some cases inflammations, which after 90 days had worsened (Lam 2003). Warheit obtained comparable results in similar investigations with rats, but the granulomae did not continue to worsen after 30 days (Warheit et al. 2004). The two studies differ not only in their laboratory animals, but also in the nanotubes used. Warheit used "laser-evaporated nanotubes," Lam used "HiPco and carbon-arc nanotubes." The differing degrees of intensity of the inflammations could be caused by the differing nanotubes, suggesting a varying impact of similar substances. Furthermore, Warheit reports that in his test 15% of the rats died from a mechanical blockage of the upper respiratory tract due to agglomerated nanotubes. This suggests the necessity of integrating the behavior of released particles into the assessment of possible negative effects.

In the project survey, Kreyling (GSF - Forschungszentrum für Umwelt und Gesundheit) suggested that nanotubes might also be carcinogenic. It is known from the toxicology of mineral fiber that incorporation (especially inhalation) of biopersistent fibers of 20 μm length and above greatly increases the risk of cancer. Because of their length of more than 50 μm and their extraordinary stability, it cannot be ruled out that nanotubes, first of all, are biopersistent (they cannot be dissolved or broken down in concentrated acids), and secondly, increase the risk of cancer after inhalation.

So far, the impact of nano-scale titanium dioxide when inhaled is also unclear. This is, first of all, a concern for the labor force producing nano-scale titanium dioxide. Colvin points out that titanium dioxide may cause inflammation in the lungs (Colvin 2003a). Rehn et al. administered a coated and an uncoated form of ultra-fine titanium dioxide (diameter < 100 nm) in single doses to rats in work-place relevant dosages (0.15, 0.3, 0.6, and 1.2 mg). They could not, however, prove inflammation and assume that both forms of titanium dioxide are inactive in the lungs (Rehn et al. 2003). On the basis of another test arrangement, Bermudez et al. found contradictory evidence. In an inhalation experiment, rats, mice and hamsters were exposed to airborne ultra-fine titanium dioxide particles (e.g. 10mg/m^3) for 13 weeks. In rats and mice they found indications for inflammation of the pulmonary tissue, which disappeared with decreasing exposure. In both species, the self-purification mechanism of the lungs appeared to be overloaded by such concentrations, which was visible in the delayed clearance of the particles. Rats showed more severe symptoms of inflammation than mice and hamsters showed none, which was interpreted to be a sign of their higher capacity for pulmonary particle clearance (Bermudez et al. 2004).

Oberdörster reported in the survey that nano-scale titanium dioxide causes pulmonary cancer in rats. High doses were applied in the test and it is not clear whether human beings are exposed to such high doses. Pulmonary cancer in rats was the result of two years' enforced inhalation of nano-scale titanium dioxide in high concentrations (10 mg/m^3 and 250 mg/m^3). Whether these results can be transferred to humans is therefore questionable.

In summary, one can state that there is the suspicion of a problematic impact of nanoparticles. Simultaneously, the experts responding to our survey confirmed that investigations were carried out under laboratory conditions administering substances of mostly very high dosages to laboratory animals. Exposure of humans to equally high concentrations is hardly imaginable except in case of severe industrial or transport accidents. A current overview of this analysis and the unanswered questions is provided by Oberdörster et al. (2005) and IOM (2005).

Impact on the environment
In addition to the above-mentioned experiments, which attempt to investigate the effects of nanomaterials on humans by means of animal testing, there are a few investigations on the environmental toxicity of nanoparticles. Generally it must be noted that the behavior of nano-scale materials and substances in the environment needs to be studied (Tomson et al. 2003).

Research on the behavior of nanomaterials in different environmental compartments and conditions is only just beginning. Researchers at Rice University are investigating the effects of nanomaterials in the soil (biopersistency, dissolution, biodegradation, aggregation, adsorption into environmental matrix) and aquatic environments (dissolution and suspension in aqueous media, sedimentation) as well as the effects of bio-accumulation (earth worms and aquatic animals) (Tomson et al. 2003).

Preliminary results of these investigations show:
- Aggregation of nanoparticles may differ in various aqueous media.
- Adsorption of foreign matter on the surface of nanoparticles is very high.
- Adsorption/desorption of organic compounds in nanoparticles could be long-term.
- Nanomaterials in natural aquatic environments could considerably change the behavior and mobility of pollutants.

Brumfiel (2003) reports that researchers at Rice University have investigated the behavior of buckyballs. They dissolved buckyballs in water, which was poured onto the ground. The buckyballs behaved with very different consequences in the ground. When buckyballs agglomerated and formed particles of micrometer size, they were absorbed by the soil like any other organic substance. When they spread without agglomeration, it was observed that water formed a kind of protective shell around the buckyballs. In this way the buckyballs may be able to pass through the soil without being absorbed and thus represent a hazard to ground water. Moreover, there are indications that nanomaterials could enter the food chain. Brumfield reports that nanoparticles might be ingested by earthworms (Brumfiel 2003).

There are concerns in research on the possible consequences of nano-scale structures for the environment with respect to the ability of nanoparticles and microparticles to take up heavy metal, radionuclides, and other hazardous substances as foreign matter. It is notable that nanoparticles would react with heavy metals and be able to transport them.

Wiesner (Rice University) undertakes research on the behavior of nanomaterials in water and has made the following statements: "Nanomaterials can move with great speeds through aquifers and soil ... nanomaterials provide a large and active surface for sorbing smaller contaminants, such as cadmium and organics. Thus, like naturally occurring colloids they could provide an avenue for rapid and long-range transport of waste in underground water" (see Colvin 2002).

One international research project concluded that nanoparticles can form in the intermixing of mining effluents and running fresh water. The nanoparticles pick up highly concentrated toxic heavy metals found in mining operations run-off. Here, too, the scientists assume that the particles might transport the poisonous metals further down the river due to chemical bonding (see Vista Verde 2002).

At the same time that scientists succeeded in making nano-scale water-soluble in order to be able to administer a medication more or less directly to the desired location, they made it possible for these particles to move freely in the groundwater.

Eva Oberdörster (Southern Methodist University, Dallas) reported on a survey in the context of her latest investigation into the impact of fullerenes on large-mouth bass and water fleas. Although the fullerenes were toxic for water fleas, the fish did not die and showed no indication of disease. However there were indications for cerebral damage and possible inflammation; furthermore, the fullerenes produced strong reactions in the fish, although the dosage applied in the investigation (1 ppm) would rarely appear in practice unless the result of an accident, the researcher reported. High concentrations of nanoparticles with unknown consequences could occur when used in suntan lotions or cosmetic products that are washed away from human bodies in water. At the same time, the behavior of nanoparticles must be taken into consideration. Presumably, individual nanoparticles would show different characteristics than agglomerated ones when inhaled or released in aqueous form.

Additionally, Oberdörster reported that not all artificially produced nanomaterials had the same effect. Her investigations revealed, for example, that fullerenes are toxic for E. coli bacteria. Coated single-walled nanotubes however are not toxic for E. coli.

Similar to the situation in the health field, it must be stated that although systematization of the field has progressed, little detailed knowledge is available with regard to actual effects in the environment. The IOM (2005:40) states that "the assessment of the environmental impact of nanomaterials, as for any other materials, will have to focus on residence time in the environment, toxicity (acute and long-term), bioaccumulation potential and persistency in living systems. Little is known about [any] of these."

Both Oberdörster et al. (2005) and US-EPA (2005) systematize the contexts and underscore that nothing is so far known about the potential impact of nanoparticles in the environment.

In a summary of research results compiled by the US Nanotechnology Initiative Funding, Dunpy Guzman et al. (2006) conclude with regard to exposure, environmental fate, and transport that: "it is unknown if engineered nanoparticles, especially those coated to reduce aggregation, will behave similarly (compared to incidental nanoparticle aerosols), the exposure assessment studies ... have focused on worker exposure, but exposure of the ecosystem and the public to nanoparticles, from either manufacturing or the use and disposal of nanoparticle-based products, needs to be quantified ... Transport studies to date have been limited to aerosol transport in the atmosphere and transport studies in porous media. However, each ecosystem component must be considered: soil, sediment, oceans, surface waters, groundwater, and the atmosphere."

Regarding toxicity, a number of studies are listed that suggest negative effects. "Many believe that surface coatings have the potential to greatly alter the toxicity, solubility, reactivity, bioavailability, and the catalytic properties of underlying nanoparticles, thus minimizing their health and environmental impacts. Unfortunately, these coatings may not persist indefinitely after release of the underlying nanoparticle into the environment" (p.1405). Studies of CdSe quantum dots with a surface coating do indicate that the coating prevents the cytotoxicity of the quantum dots; however, the stability of the coatings is unclear. Finally the authors also point to the potential global impact, particularly the atmospheric impact, referring to studies which show that nanoparticles are key components in many biogeochemical processes.

Summarizing the current status of knowledge it must be stated that with respect to the possible effects of the nanoparticle even the most recent surveys tend to look like catalogues of questions and that little verified knowledge exists. This also holds true for problem studies such as those undertaken by E. Oberdörster (IOM points out that the effects found by E. Oberdörster are not necessarily consequences of nanoparticles) as well as for "all-clear" messages, which likewise are frequently based on single studies.

4.6.2 Nanoscalar titanium dioxide in suntan lotions

The use of nano-scale titanium dioxide in suntan lotions addressed in the following case study. Titanium dioxide is increasingly used as an effective protective substance against ultraviolet radiation in several cosmetic products. Nano-scale titanium dioxide was chosen for this case study essentially for two reasons. First, because it involves a non-contained application on a large scale. Secondly, because it is the application of the substance which highlights the problem of assessing the effects of new nanomaterials.

The inclusion of nano-scale titanium dioxide in suntan lotions is scientifically disputed. Titanium dioxide in its micro-size has a long tradition as a component in suntan lotions. Onlu more recently have manufacturers been producing and adding nano-scale titanium dioxide to suntan lotions.

As we know now, a nano-scale substance can have completely different characteristics (chemical, optical, etc.) than its micro-scale counterpart. Nano-scale titanium dioxide is transparent, while micro-scale titanium dioxide is white and therefore is used as a pigment, for example, in wall paint. Nevertheless nano-scale titanium dioxide was classified as equal to micro-scale titanium dioxide regarding effects by the respective regulatory authorities in the USA (Food and Drug Administration 1999) and the EU (SCCNFP 2003).

The classification and regulation of nano-scale titanium dioxide and its impact on the human organism can serve as an example for the discussion processes concerned with nanoparticles and their hazards and risks. Nano-scale titanium dioxide was also chosen for this case study because it is at the center of the arguments brought forward by ETC Group against nanotechnology. The NGO is demanding a worldwide moratorium on nanotechnology research. Using the example of nano-scale titanium dioxide, the NGO underscores the repeatedly expressed considerations of scientists with respect to possible hazards and lack of knowledge about the environmental behavior of nanoparticles. At the same time, the NGO criticizes the regulatory equation of macro-scale and nano-scale titanium dioxide, which does not sufficiently consider their differing behavioral characteristics.

Titanium dioxide belongs to the group of physical light protective filters (also called inorganic or mineral filters). These are metal oxides, which filter the UV light predominantly by reflection and diffusion. Major representatives are TiO_2 and ZnO. They are biologically and chemically stable and very rarely cause irritations and phototoxic or photo-allergic reactions. With their high capacity to absorb ultraviolet radiation, nano-scale titanium dioxide and zinc oxide are ideal for use in cosmetics.

4.6.2.1 Hazards of nano-scale titanium dioxide

The cosmetics industry defends its use of nano-scale titanium dioxide in suntan lotions with own investigations demonstrating the innocuousness of titanium dioxide. The safety of nano-scale titanium dioxide was also confirmed by the Scientific Committee on Cosmetic Products and Non-food Products Intended for Consumers (SCCNFP), an EU body, in October 2000. The committee decided that along with its conventional counterpart, nano-scale titanium dioxide could also be included in the list of approved UV filters: "The SCCNFP is of the opinion that titanium dioxide is safe for use in cosmetic products at a maximum concentration of 25% in order to protect the skin from certain harmful effects of UV radiation. This opinion concerns crystalline titanium dioxide, whether or not subjected to various treatments (coating, doping, etc.), irrespective of particle size, provided only that such treatments do not compromise the safety of the product" (Europäische Kommission 2004). The SCCNFP decision was based however solely on industrial studies not accessible to the public.

In the application of suntan lotions containing nano-scale titanium dioxide, the question arises as to whether it can reach living cells, enter, and either damage them or migrate even further into the body. Moreover, it is not known whether titanium dioxide particles can transport other undesirable substances into the body. Colvin assumes that the risk from titanium dioxide risk is much lower than that of sun bathing itself (Colvin 2003a).

Scientific investigations on coated nano-scale titanium dioxide report that when applied to the skin, the majority of the titanium dioxide remains localized in the upper layers of the skin (stratum corneum). The concentration of titanium dioxide in the upper layers of the stratum corneum decreases drastically over the time, as the skin regenerates (exfoliation). In the deeper layers of the skin, however, the concentration of titanium dioxide particles decreases much more slowly over time (Rickmeyer 2002).

Pflückner reports that coated nano-scale titanium dioxide can be found in hair follicles, which are found in the deeper layers of the skin. However, there are no indications for their movement from the follicles into living cells (Pflücker et al. 2001). This confirms the results of earlier investigations (Lademann et al. 1999). Rickmeyer reports that only very low concentrations of coated titanium dioxide were found in the tissue surrounding the follicles. Contact with living cells "exists only in an evanescently minimum scope" (Rickmeyer 2002). Bennat/Müller-Groymann report that nano-scale titanium dioxide in fatty solutions can penetrate the skin better than in aqueous solutions and do not rule out the possibility of penetration of particles into deeper layers of the skin (Bennat/ Müller-Groymann 2000).

However, Tinkle et al. report that beryllium particles of one-half to one micrometer in size rubbed into the skin have reached the epidermis and in rarer cases the deeper dermis thus bringing them into contact with living cells (Tinkle et al. 2003). Admittedly this result does not say anything about the impact of titanium dioxide on the skin, but it makes clear that in addition to particle size substance class must also be considered in judging possible effects.

The safety of nano-scale titanium dioxide is also being called into question elsewhere. A report by the Royal Society of Engineering e.g. suggests that with the reduction of particle size the number of free radicals on the surface of titanium dioxide increases, possibly leading to skin damage

(Royal Academy of Engineering 2003). This view is backed by an investigation by Dunford et al. and Uchino et al., who report that when titanium dioxide is exposed to sunlight, in vitro as well as with human cells, the cell's DNA is damaged by photocatalysis (Dunford et al. 1997, Uchino et al. 2002, Colvin 2003a). Rickmeyer assumes that these photocatalytic effects may be reduced drastically by coating, but cannot be entirely avoided (Rickmeyer 2002). Moreover it is known that nano-scale titanium dioxide causes cell death in the fetuses of Syrian hamsters (Rahman et al. 2002).

Butz (2006) reports on the results of the Nanoderm research project being promoted by the EU, which is specifically investigating the absorption of titanium dioxide by the skin. His conclusion is that, normally, penetration is restricted to the stratum corneum disjunctum, occasionally Ti is found in s.c. compactum, Ti is rarely detected in the stratum granulosum, Ti is detected in the stratum spinosum in most cases, and Ti spots in the dermis are identified as contaminations.

Moreover Butz states: "To our surprise, the particle shape had no influence, it appears that TiO2 particles are mechanically rubbed into the horny layer / hair follicles / furrows without diffusive transport and thus far there is very limited exposure to vital tissue, but there are open questions: particles in the 1–2 nm range might behave like small macromolecules and penetrate; transglandular pathway clearance from follicles (+ glands?)"

All in all, these research results suggest that nano-structured titanium dioxide as used in the cosmetics industry does not penetrate into the deeper layers when used on healthy skin and that possible negative effects on the cells do not represent a problem of the first priority. Regarding potential effects within the individual environmental compartments, however, little is known.

4.6.2.2 Expert survey

In the course of the expert survey, a number of new as yet unconsidered aspects emerged with respect to this case study. Asked about possible negative effects of nano-scale titanium dioxide, the experts gave the following answers: The majority was of the opinion that there are no indications for the penetration of nano-scale titanium dioxide through all layers of the skin. Only one expert was generally more skeptical about the problem, but without providing any new information. They all agreed on the following problem: No investigations have looked at application to wounds or persons with dermal allergies. And there are no investigations on application to inflamed and (sun) burnt skin. In both cases there is the possibility of nano-scale titanium dioxide coming into contact with blood and/or living cells. Studies on the persistence of particles in the body, particularly in babies and children, are required. Therefore, precaution is still necessary.. One expert noted: Nano-scale titanium dioxide can easily penetrate the skin when coated with electrophilic, oily molecules.

4.6.2.3 Summary

The case study on the use of nano-scale titanium dioxide in suntan lotions illustrates the problems associated with future applications of nanoparticles. A number of experts question the full safety of the application. At the same time, a number of important investigations on the effects of nano-scale titanium dioxide in the human body are lacking, although the main use, i.e. application to the healthy skin, appears to be unproblematic, according to our latest knowledge. However, the fact that this is a non-contained, environmentally open application could cause a number of problems. We do not know how the structures/substances behave in nature after being washed off or leaving the skin in the course of natural skin exfoliation. In light of already occurring problems with suntan lotion residues on lake and sea shores, these questions should not be neglected.

Another problem is that nano-scale titanium dioxide is already being produced in large quantities, yet no knowledge is available on potential hazards in the production plants. The question of the effect of coatings on nano-scale titanium dioxide also remains unanswered. Assuming they really prevent catalytic effects, it is not known how long the coatings remain stable and how they react in surroundings other than on the human skin.

The EU's equal treatment of micro-scale and nano-scale titanium dioxide is at the very least problematic in its lack of consideration of possible consequential environmental effects. Furthermore, publication of the related industry studies should increase the trust in the EU decision. At the same time, it is necessary to pursue unanswered questions regarding the safety of particles. With this in mind, the initial avoidance of non-contained applications, a reasonable measure based on the precautionary principle, should only be discontinued when, as in the case of nano titanium dioxide, essential, fundamental knowledge is available.

4.6.3 Life cycle analysis of nanomaterials

As already remarked, possible harm by nanoparticles may occur especially in environmentally open applications. A look at the life cycle of nanomaterials reveals several points where such a release may take place:

1. Nanomaterials production processes vary greatly. Industrially manufactured nanomaterials are not generally produced by combustion processes (with the exception of CVD/DVD and flame-assisted deposition), but mainly in liquid or closed gas-phase reactors. Therefore, direct exposure to nanomaterials ought to be limited. Exposure in the work place and research laboratory and possible exposure of man and environment due to accidents are all problematic.
2. Products: Most nanoparticles are contained or immobilized in products. Examples include nanotubes in video monitors or particles in paint coatings. The probability of release is presumably low.
3. Disposal and recycling: Behavior during disposal and recycling has not yet been extensively investigated, but release of individual nanoparticles is presumed to be limited. However, one must recognize that only very preliminary knowledge exists concerning this problem.

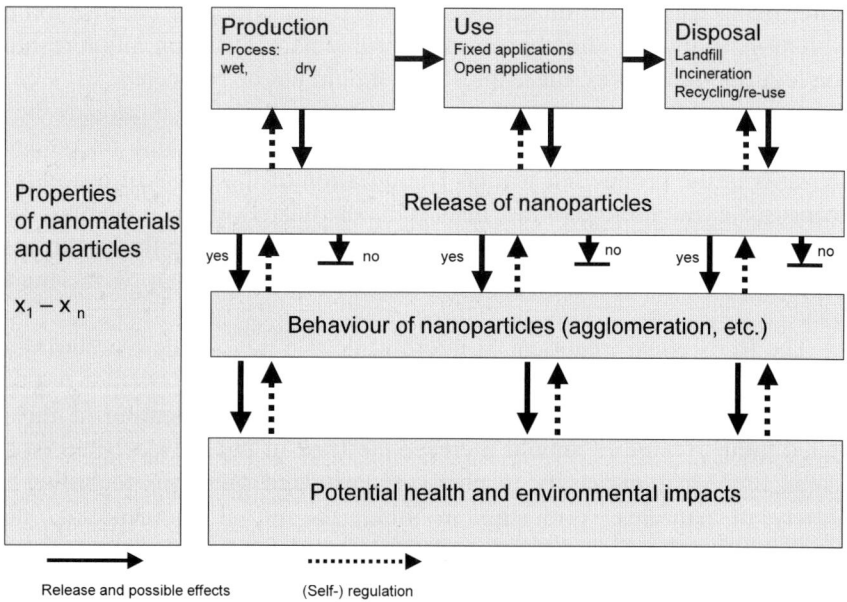

Fig. 48. Potential release pathways of nanoparticles[96]

In the course of the survey, most experts pointed out possible risks that occur during the life cycle. Exposure at the work place, accidents during manufacture and transport, and environmental pollution via industrial wastes are mentioned as possible sources. In the case of possible contami-

---

[96] Source: Haum et al. (2004)

nations due to accidents, it is entirely unclear, so far, what impact the individual nanomaterials have on affected ecosystems and humans. The problems will become worse if additional consumer goods are produced without thorough testing or even no testing at all in the future, as is the case in the cosmetics industry. Closely connected to the question of the effects of accidents, waste, and disposal is the issue of biopersistency. It is necessary for the effectiveness of some nano-applications, such as those that must release their effects in a specific part of the body. However, sufficient knowledge for an assessment of long-term effects is not yet available.

One expert remarked in this context that a great deal of future production of nanomaterials will take place under cleanroom conditions, which will reduce the exposure of production staff. Presently, it is absolutely unclear whether and to what extent workers in plants or researchers in laboratories are exposed to particles. Industrial accidents still represent a potential risk. Likewise, one expert pointed out that due to the currently still small volume of nanoparticles being produced, the sum risk of exposures in the work place is rather small.

In the field of medical applications, the experts assumed that existing test procedures would reveal potential risks, before mass production takes place. On the basis of the differing approval procedures in specific fields, the question arose as to whether the same would be guaranteed, for example, in the field of consumer goods.

In summary, one can conclude that:
- most manufacturing takes place in contained systems
- Particles are often securely integrated in products and thus risks are limited
- the problem of disposal and recycling is unsolved
- intended and unintended release is problematic

These statements probably must be amended for new and modified production systems.

4.6.4 Discussion

The behavior of nanoparticles differs from that of structures at the macro-level; this also holds true for identical substances. Some of the studies discussed above document a surprising behavior. As shown, there are numerous reasons for concern as well as clear indications of toxic effects of nanomaterials, particularly nanoparticles, on the environment and human health. Nanotubes and buckyballs could be of special significance in this respect.

The knowledge we have is all preliminary, in part even contradictory, and deals with only a fraction of all possible effects. At the same time, the transferability of the knowledge gained so far appears to be low. The survey of the experts revealed that current knowledge is much too incomplete to form the basis for a comprehensive risk assessment or to implement evidence-based risk management. For this reason measures based on the precautionary principle are required.

Asked whether general statements can be made about the toxicity of nanoparticles, the experts' responses were mainly negative. It is too early to classify nanomaterials into groups and categories that could be characterized with respect to their negative effects. Classification is only possible when specific nanoparticle properties responsible for toxic effects are identified. The statement that smaller particles are, as a rule, more toxic than larger ones, as their surface-volume ratio is greater was repeatedly made. Admittedly, the chemical composition and physical structure must always be considered. One expert even limited this rule to those particles having low solubility. One expert proposed a classification according to:
- Biopersistency
- Biologically accessible surface area
- Aggregation behavior in various environments

This type of classification seems to encompass the greatest known potential hazards and thus, with the inclusion of mobility in environmental compartments and in the human body, could become a guideline for future research on risks.

In addition to the uncertainty about possible negative effects, it can also be stated that the occurrence of negative effects due to manufactured particles presently appears to be relatively rare. As a consequence, continued research on the behavior and potential effects of nanoparticles in case of a release is required. Since general statements are not currently possible, there is likewise an urgent demand for classification.

4.6 Case study 5: The risk potential of nano-scale structures

Several experts remarked in the course of the survey that the possibilities offered by nano-scale structures and systems are enormous. To suspend development efforts now already on the basis of a profound suspicion of possible toxicity, would be far out of proportion in relation to the expected gains from nanotechnology. Instead, attention should be given now to establishing scientifically based criteria for the evaluation of risk and risk management.

Since research on the risks of nanoparticles is still in its infancy, only very general advice for handling nanomaterials can be given:
- Differences in size, surface structure, and chemical composition require that the possible effects on human health and environment for each individual nanomaterial be investigated.
- There is a great need of classification.
- Non-contained applications of some nanoparticles and nanostructures should be avoided until potential environmental effects have been addressed. This holds true for the majority of nanoparticles at present.
- The behavior of nanomaterials released during disposal should also be investigated.
- Biodegradability and increased tendencies for agglomeration could be ways to minimize or to avoid environmental hazards (ecotoxicity).

In this context the development of a strategic research and development concept is important. It should provide the necessary means and also maintain a balance between technological development and understanding of effects. This would mean a better intercoupling of both research directions. One expert proposed a comprehensive concept for the investigation and assessment of possible hazards:

"In my opinion, the most urgent need for action with respect to regulation is in the development of a new, strategic concept for the toxicological evaluation not only of the sites and organs where absorption occurs, but also the secondary target organs. This concept should be based on modern, genomic, proteomic and toxiconomic investigations with high-throughput technology. This concept would not only allow comprehensive evaluation and specific regulation of a new product, but would provide the manufacturer, in connection with a suitably equipped toxicological laboratory, with a prompt risk assessment. Such a strategic concept should be developed in interaction between research, manufacturers, and the regulatory authorities. In this way Germany could extend its leading position in the field of sustainable development of new nanoproducts."

Basically it seems important, on the one hand, to understand the basics of the effect of nanoparticles; on the other hand, more research must be carried out in those areas where production quantities are foreseeable. References from the experts as to contamination resulting from accidents in production and transport underscore the urgency. A further research focus can be found in answers given to the question as to whether certain nanoparticles or structures are intended for use in open applications; if so, their potential impact must be more closely investigated beforehand.

5 Formative approaches to a sustainable nanotechnology

It is not – or only to a very limited extent – possible to steer technological development by means of political intervention; it is rather the interaction of the various players, which usually leads to a developmental path that can be concurrently shaped "en route." Leitbilder can also be useful as tools for directing and shaping development. Inasmuch as nanotechnology is for the most part still in an early phase of development, there still exists, at least in principle, a large degree of freedom, thus allowing research efforts to be steered towards "sustainable development."

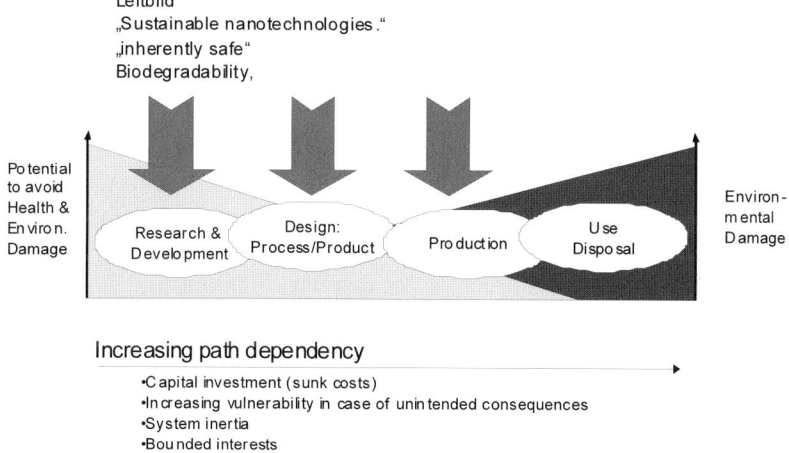

Fig. 49. Windows of opportunity for shaping technological development in the product life cycle[97]

The illustration demonstrates that throughout the entire process – from basic and applied research through development, use, and disposal phases – appropriate precautionary options can and must be developed for each phase. Once options are developed, they can also accordingly be applied to

---

[97] Source: Rejeski et al. (2003)

subsequent research. Various players are involved and responsible in each of the phases (in some cases collectively). The development of precautionary options can already begin in the basic research phase (research into the consequences of scientific development), the results of which can subsequently lead to research and development efforts in the area of applied research.[98] In our view, the early stages offer the greatest opportunity, relatively speaking, to avoid potential environmental and health risks.

The most important aspect of including precautionary aspects in early research phases is that knowledge about hazards can already be recorded and in turn incorporated into decisions about the direction of further research. Of interest for the field of nanotechnology are the "foresight guidelines" published by the Foresight Institute (2001). However, it is important that above and beyond the extreme cases addressed by Foresight, further aspects also must be considered, which, like the Foresight guidelines, would provide direction for development in the sense of "sustainability guidelines."[99]

The design development phases for production process and product design are, in a sense, already predetermined inasmuch as their foundation is established in the initial research and development. Yet there is still a relatively large degree of freedom in process and product design, which can be decisive for aspects such as the "intrinsic safety" of the technology, processes, and products, as well as the impact of later phases. Design options are more limited, but to a large extent still possible.

Options for shaping design development become significantly fewer in the production phase and beyond. But often it is still possible to implement additional measures by means of product safety sheets, through the hazardous substances categories that regulate the handling of processes and products, and through the enforcement of disposal regulations.

Moreover, each of the subsequent phases has its own players capable of influencing development. Each has a share in the development and corresponding opportunities for influence and thus contributes decisively to potential impacts of the technologies, processes, and products being developed. Management systems can be of great importance in this respect. To implement comprehensive quality management and integrate safety,

---

[98] This is a simplified observation; a distinct separation between fundamental research and applied research is not possible, but a differentiation based on the respective primary goals is.

[99] Since sustainability aims for safety barriers to avoid non-sustainability but at the same time represents a complex *Leitbild,* addressing many more aspects than just risks, simple guidelines may not be sufficient; see the discussion of the guiding model nanobionics.

health, and environmental aspects throughout the entire value chain right from the start, comprehensive communication channels are essential. However, such a flow of communication between the various players often does not exist unless large-scale enterprises with their own comprehensive management systems or firms situated between two successive production stages are involved. The communication between those who may be affected at the end of the product cycle and those at the beginning is frequently non-existent or only developed to a very limited extent.

Principally, it is necessary to close these communication gaps; this is particularly important with respect to new products and substances for which little information is available and those, as in the case of nanoparticles, which are not addressed by current risk assessment processes and therefore may not be caught by current regulatory screening processes.

One important aspect is improving the flexibility and accessibility of frequently closed communication systems, as well as ensuring that they are capable of handling challenges and uncertainties in a responsible manner.

In addition to the scientific and technical development pathway, path dependencies must also be taken into consideration. Path dependencies are influenced by investment in infrastructure and production capacities, as well as the know-how and knowledge invested in particular lines of technology, processes, and products. Here, too, in the course of the product cycle the development pathway becomes increasingly more rigid (sunk costs).

Path depencies created through economic activity and certain investments in knowledge might create undesired side effects not containable by individual actors. Reversing unintended technology effects can better be achieved if various institutions and groups of players act cooperatively. Coordination and cooperation of actors is, in principle, a governance problem in which the influence of the various players in the design process often cannot be clearly identified. As it has been shown, the consequences of nanotechnological innovations are only predictable to a limited extent. The political players' opportunities for steering development are limited by the individual dynamics of the social and/or corporate subsystems. It is therefore important that both subsystems are made more flexible, so that they remain open and accessible and are able to react to uncertainties (including those of a normative nature) in a problem-solving manner.

It is with this perspective in mind that specific design approaches – both of a joint as well as player-specific nature – should be developed.

5.1 Cooperative design approaches

Cooperative approaches are those requiring the cooperation of more than one player. In the following, we look at the opportunities for influence offered by the development of Leitbilder (of varying scope), technological road maps, and new methods of engineering results assessment and outline their potential for shaping technological development. Road maps, such as those used in the fields of information and communications technology, are jointly produced by firms within a specific industry during a phase of pre-competitive cooperation. Road maps contribute to building consensus, a prerequisite for standardization, and thereby allow for greater confidence in planning. A more refined appreciation of quality today means that it is no longer unusual for such road maps to also address safety and protection issues involving end users, public health, workplace, and environmental concerns. As a rule, road maps are often strongly influenced by the possibilities a new technology has to offer (technology push innovation). Methods and processes for developing an effective Leitbild are, in contrast, more rigorously geared towards developing a positive model for development, one capable of influencing the direction of development, i.e., moving toward "sustainable" technological development (demand/problem pull innovation).

The Leitbild plays an important role in the evolution and stabilization of scientific paradigms and technological pathways for development; this has been shown by research into science and the genesis of technology and innovation (see Kuhn 1975, Dosi 1982, Bijker et al. 1987, Dierkes et al. 1992, Dierkes 1997, Mambrey et al. 1995, Hellige 1996, David 1985, 2000). Particularly in the nanotech-science field, the great importance placed on Leitbilder and visualizations right from the very beginning was, and is, incalculable and has been the object of scientific investigation for some time already (see Baird 2003, Robinson 2003, 2004, Nordmann 2003, 2004).

Whether and to what extent such models can be deliberately and successfully implemented as tools of influence in shaping technological development is still contested (see, for example, Mambrey et al. 1995, Hellige 1996, Meyer-Krahmer 1997). In corporate strategic management, one often finds a comparably successful deployment of various "management by" approaches, including "management by guiding principles," in other words management with the help of corporate Leitbilder (see Bea, Haas 1995, Blättel-Mink 2001, KPMG 1999). However, questions regarding the selective implementation of Leitbilder can only be partially answered at the theoretical level. Systematic empirical investigations combined with experimental field trials would be necessary for a full explanation.

To be able to exert influence and shape development with the help of a Leitbild, one must understand the prerequisites for their effectiveness and how a successful Leitbild functions. It can motivate the formation of a group identity that serves to coordinate and synchronize the activities of individual players, reduce complexity, and structure perception. Among the most important prerequisites for its effectiveness are therefore vivid imagery and emotionality, a successful "role model" function, as well as an ability to offer a balanced approach to wishes and feasibility. In short, the success of a Leitbild depends on its degree of resonance in the minds of the players. Vivid imagery is important for clarity and reduction of complexity. Above and beyond its emotional content and the values it conveys, the Leitbild also motivates actors and provides direction. "Closed-loop economy" is an example of an effective technology-oriented Leitbild at a mid-level of operationalization. Bionics or biomimetics, too, (nature's 'technological' solutions as a role model) and „green chemistry" have proven to be effective.[100]

5.1.1 Various Leitbilder for sustainable nanotechnology development

A technology-oriented approach has been chosen for the development of Leitbilder for sustainable nanotechnology. Starting point of this approach is the examination of the technology regarding possible contributions to sustainable development, followed by a consideration of the associated prospects and risks with a view to the goal of sustainability. In the current phase, in which nanotechnology is undergoing above all a technology-driven development dynamic (technology-push innovation), such an approach may well be reasonable and promising. Alternatives or competitors would include the problem-oriented and the needs-oriented approaches. In the problem-oriented approach, the focus would be on climate protection, resource conservation, and risk minimization, for example, and nanotechnology would only come into play in those areas where it could potentially contribute to the solution of anticipated problems. The same is true for the needs assessment approach; for example, for the goal of "sustainable construction and housing," nanotechnology might be asked to contribute a possible solution. Enhancements to the selected technology-oriented approaches in the area of needs assessment and problem-orientation are pos-

---

[100] Cf. Ahrens; von Gleich 2002

sible by focusing on specific areas of application, i.e., on sustainable applications and implementations of nanotechnology, such as the construction of lightweight (recyclable) vehicles utilizing nanofibre-reinforced materials.

The requirements for a successful Leitbild for sustainable nanotechnology, already described in greater detail in Section 2.3, are as follows: It should be graphic and clear; it should provide motivation by conveying values and aims; and it should build upon the true capabilities of nanotechnology, that is, it should maintain a realistic approach regarding what actually can be achieved with nanotechnology. It is helpful for a somewhat more systematic approach to distinguish according to the time-frames of such a Leitbild with respect to goals as well as to the technical potential and feasibility. A consideration according to time frames could begin with defensive and possibly short-term realizable potentials for risk minimization and damage avoidance and lead to mid-term development and planning perspectives, such as are formulated in the technological road maps of far-sighted businesses and branches, as well as to a long-term, visionary plan.

In the short-term time frame, it is possible to derive the Leitbild "Resource-efficient Nanotechnology" from the current debates. For the mid-term, the Leitbild "Consistent[101] and Intrinsically Safe Nanotechnology" is possible, while "Nanobionics" represents an adequate Leitbild for a long-term perspective. These Leitbilder represent varying scopes and strategies.[102] They build on one another – the long-term Leitbild incorporates the goals of the short-term one. Nanobionics should therefore satisfy the requirements of resource efficiency, consistency, and intrinsic safety.

The development of such Leitbilder with the idea of influencing technological development can, as a rule, only be successfully realized as a further explication (differentiation and concretization) of already present and operative Leitbilder. In exceptional cases, the successful development is

---

[101] "Consistent" should be understood here as the qualitative and quantitative embedding of socio-technical metabolism in the natural (Huber 2001). This can be achieved through an opening up of the material and energy flows between technosphere and ecosphere (for example, by transition to regenerative material and energy sources and attention to the biological or photochemical biodegradability of substances), but also through a particularly effective closing off between technosphere and ecosphere. This can be realized by means of "closed" applications and best containment practices and by the most effective and high-quality recycling (see for example McDonough & Braungart 2002).

[102] For example, the distinctions between follow-up environmental protection (end-of-pipe), integrated environmental protection, and sustainable development and between strategies of efficiency and consistency.

perhaps possible as a mixture of the creative with the abstract that is derived from the already knowable, i.e., in an empirical-synthetic interrelationship based on dialogue and communication. That means that the Leitbild's main concepts must tie into current debate and already effective directional statements and guidelines. At the same time, substantial differences between the new Leitbild and existing approaches can (and likely must) come to light. Leitbilder that have been successful in the past have usually provided significant elements of surprise and irritation. In this respect, one must also confront the danger, well-known from the sustainability debate, that real-life goal conflicts and conflicts of interest will be concealed instead of resolved at the Leitbild level.

The "resource-efficient nanotechnology" Leitbild focuses on the product life cycle. The potential environmental and resource relief in the use phase, for example, should not be thwarted by excessive production costs or by recycling problems or waste disposal. Environmentally friendly manufacturing methods therefore are also a part of the Leitbild.[103] Nano-environmental engineering, i.e., "end-of-the-pipe implementation" in the waste-gas and waste-water treatment must also be integrated here.[104]

Essential elements of a "consistent and intrinsically safe nanotechnology" would be, for example, the deployment whenever possible of "intrinsically safe substances and structures" (for example, those that are rapidly biodegradable, or with a tendency to agglomeration and therefore to the loss of their specific properties at the nanoscale level) in the course of an "intrinsically safe technology" (i. e., aqueous manufacturing method for nanoparticles instead of open, dust-producing methods), within the framework of an "intrinsically safe application system" (i. e. closed system instead of environmentally open applications, nanoscale structures preferably embedded in a matrix). An uncontrolled spread of nanostructures could, for example, be prevented by utilizing the following design and selection principles:

1. More rapid dissipation where possible of problematic "nano-properties" (for example, by means of agglomeration)
2. Rapid metabolism of the utilized substances (biological or photochemical biodegradability)

---

[103] For example, process-integrated environmental protection, "green chemistry" and "clean nanotechnology."

[104] A large number of the 62 lectures that were given at the 227th ACS National Meeting in Anaheim, 28 March – 1 April 2004, on the subject of "nanotechnology and the environment," represent projects in which the contribution of nanotechnology to environmental protection is emphasized (follow-up as well as integrated measures), cf. http://oasys2.confex.com/acs/227nm/techprogram/.

3. Negligible bioavailability and bioaccumulation of substances and particles (for example, by means of minimal solubility or liposolubility)
4. Restriction to closed applications (avoidance of environmentally open applications, high containment)
5. Avoidance of possible self-replication[105]

In the most far-reaching visions, which have accompanied the development of nanotechnology from its very beginning, two comparatively divergent lines can be distinguished. There is the much cited vision of absolute control, i.e., controlled technical development, "atom by atom." This is a matter of extending control and precision approaches derived from the macro-world beyond the micro-world (semiconductor technology, micro system technology) into the nano-realm. This mechanistic top-down approach contrasts with the biological bottom-up approach inspired in particular by biology and system theory. In this approach, the emphasis is on the self-organizing capabilities of (living) nature. By means of structural (hierarchical) design at different levels, beginning from the atomic level, nanotechniques can revolutionize material properties and applications. Thus, principles of biological development that have proven themselves in the natural world for millions of years (e. i., in bones and shells) can be applied in order to allow materials and components to "grow" (e. i., by template-controlled crystallization). If we can succeed in connecting those object-inherent tendencies and capabilities of self organization, a significant part of the energy, material, monitoring, and control expenses that are necessarily associated with the mechanistic vision could be avoided. The central element of the Leitbild nanobionics can thus be formulated as the utilization of and collaboration with principles of self-organization in nature.[106] And this is not limited to the organic. Induced and directed crystal-

---

[105] There exist proposals and guidelines addressing all aspects in the nanotechnology debate. Before the discussion initiated by Joy, Merkle had already proposed an "inherently safe" architecture for self-replicating nanobots (broadcast architecture, see Merkle 1994). Also interesting in this respect are the "development principles" and "design guidelines" in the "Foresight Guidelines on Molecular Nanotechnology" (Foresight Institute 2001). To avoid proliferation and misuse they call for: non-replicability in the natural environment, detectability, reliance on a single non-naturally occurring "fuel," reliance on remote control signaling (broadcast), programmed self-destruction, and avoidance of mutation and evolution.

[106] The Leitbild nanobionics, with its fusion of nanotechnology and bionics, is not our invention. It is a result of the new opportunities that have opened up for biomimetics by progress in nanotechnologies. For example, there have been two Heraeus seminars on nanobionics in the recent past (cf. www.nanobionics2.de).

lization or template-controlled structure formations are among the most fascinating examples in the inorganic field (see, for example, Singh et al. 1995; Niesen, Aldinger 2001). Techniques that build on the capacity of molecules for directed reaction, alignment, movement, and combination and on the capacity of cells, organisms, and eco-systems for self-organization and homeostasis are among the most promising perspectives of nanotechnology, notably when considered from the point of view of sustainability.[107] Self-organization effects comprise – together with the exergy source sunlight – the fundamental principles of evolution and ontogeny. In the nanobionics Leitbild, the, in some respects very old idea of an orientation towards the "cooperative productivity of nature with the productivity of man in an allied technology"[108] comes together with the longtime fascination of bionics and the new nanotechnical possibilities.

The Leitbild nanobionics originated in the considerable overlap between bionics approaches at the molecular level (for example, in the areas of material bionics and biomineralization) and nanotechnology approaches, which in the framework of the more biological "bottom-up paradigm," focus on the principles of self-organization of molecules (cf. ZTC VDI/TZ 2003).

The term bionics, or biomimetics, is currently among those lines of research that not only engage enthusiastic, dedicated players who are passionate about their field but also arouse fascination and enthusiasm in the public, particularly among younger people. This fascination is largely based on the bionics Leitbild: a promise of "ecological adaptiveness," "evolutionary approvedness," and the "fit of solutions within natural contexts" on the basis of the metaphor "learning from nature" combined with the fascinating achievements of organisms and the cleverness and often the "elegance" of their "technical" solutions (von Gleich 2001). Ecological suitability, resource efficiency, and risk minimization are all demands that will be made of technical innovations as we move along the path towards sustainability. Even when technical innovation alone will not suffice in bringing us closer to this goal, it should be clear that technical innovations (and this certainly includes "basic innovations") will play an important

---

[107] Cf., particularly, some results of the 'Technology Analysis of Nanotechnology' conducted by the 'Technology Centre' of the German Association of Engineers (VDI Technologiezentrum 2002, p. 105f). The center has singled out bottom-up construction of material structures, in the sense of a technical usage of self-organizing phenomena, as especially promising and is preparing a monitoring report specifically for technology early recognition (Future Technologies Consulting / VDI-Technologiezentrum 2003).

[108] See, for example, Ernst Bloch 1973 p. 807ff..

role in the course of sustainability strategies (cf. Federal Government of Germany 2002, Huber 2004).

At the same time, it must again be emphasized that Leitbilder such as "Nanobionics" are intended to provide direction but no guarantee of environmental compatibility. Leitbilder should not be misunderstood as "labels." The developments that are realized in the pursuit of such Leitbilder must, of course, still undergo a comprehensive assessment process. Direction alone does not guarantee success, although direction makes success a little more likely compared to technological developments in which environmentally benign aspects are attached to technologies, processes, and products after they have come into existance.

Scientists, end-users, firms, and policy makers should actively pursue the positive goal of sustainability from the very beginning of technology development with the help of Leitbilder such as "resource-efficient", "consistent, and intrinsically safe nanotechnologies" or "nanobionics". Demand for resources should be reduced and provision should be made to minimize or avoid potential risks. One can expect that at least the probability of achieving these objectives is increased by a conscious effort. However, one cannot expect any certainty regarding the achievement of objectives. Leitbilder, viewed in this way, are only an intention or a promise, and their redemption or redeemability must repeatedly be re-assessed. Neither bionics nor nanotechnology nor nanobionics are per se more sustainable than other fields of technology (von Gleich 2001). It is therefore necessary to assess the promises conveyed by the nanobionics Leitbild with respect to naturalness, suitability (and/or fitability), evolutionary approvedness (mimimization of risk), ingenuity, and even elegance and compare these to actual content and feasibility. This can occur qualitatively by means of a prospective TA, quantitatively by means of a prospective life cycle assessment, or process-concurrently, that is, using all the methodological approaches developed and introduced in this study.

It is only possible to touch on the already mentioned three Leitbilder here. A first general impression of some of their central elements can be found in Table 40. The brief illustrations that follow with the examples of spider silk and biomimetic synthesis may be more helpful than further definitions. It is rather unlikely that further explications alone can successfully take place at the much-cited "green table." The testing and safeguarding of the resonance of such guiding models can only take place in the course of public discourse. Leitbilder are not simply "created"; they are also not simply oriented to the lowest common denominator or awareness available. A successful Leitbild also needs imagination as well as a healthy portion of irritation and provocation. Particularly irritation and provocation can trigger an exceptionally clear and far-ranging resonance.

Table 40. Leitbilder for sustainable nanotechnology

Leitbild	Issue / Slogan	Focus	Examples
Resource-efficient nano-technologies	Environmental relief, environmental engineeringReduced resource load with high yieldLower consumptionLess damage	Quantity of energy and material flows (over life cycle) relative to social benefit	Wear-resistant and low-friction surfaces (mechanical engineering), highly specialized membranes (biotechnology, fuel cells)
Consistent and intrinsically safe nanotechnologies	Health impact and accommodation in metabolic principles and capacities of nature, minimal depth of intervention, failure friendliness	Quality (and quantity) of material and energy flows with respect to environment, health, and technical risks	Rapid biodegradability or closed applicationsRecyclable nanotubes in lightweight construction, spider silk
Nanobionics	Technology following nature's example, life-supporting, in cooperation with the self-organization principles of nature	Quality of the technology (as a form of interaction with nature)	(Bio)catalysts, enzyme technology in chemical reactions, biomimetic synthesis

Elements of these Leitbilder can and should serve right from the start of research as directions for the development of a nanotechnology associated with great hopes and expectations.[109] They are expressions of a desirable development and are also grounded in practical reality. Thus two fundamental preconditions for successful Leitbilder are satisfied. Exactly this is what makes the following two examples so interesting: spider silk as an illustration of how nature has managed to develop a fascinating, "consistent and intrinsically safe nanotechnology," and biomimetic synthesis as a "nanobionic" example that may well no longer be so far from technical realization.

---

[109] Cf., for example, (Drexler 1986; Smalley 1999)

5.1.1.1 Spider silk: consistent and intrinsically safe nanotechnologies

The threads produced by spiders show huge diversity and various combinations of properties making them a fantastic materials group. With up to seven different spinnerets, many spiders are capable of producing threads with the most varied properties: for capture (extensibility, tackiness), conservation of prey, self-protection (defensibility), and as an aid in movement, etc. (see Fig. 50). A glance at the molecular structure of the silk threads reveals how spiders have "nanotechnically" managed to integrate ductility (stretchability) and tensile strength in order to successfully capture their prey.

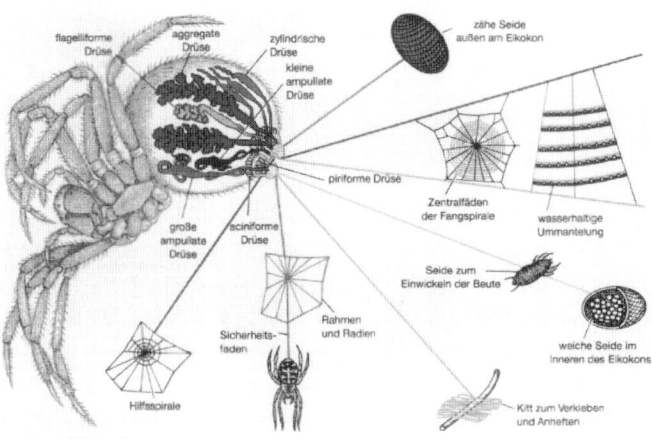

Fig. 50. Spider silk[110]

---

[110] Source: (Vollrath 1992)

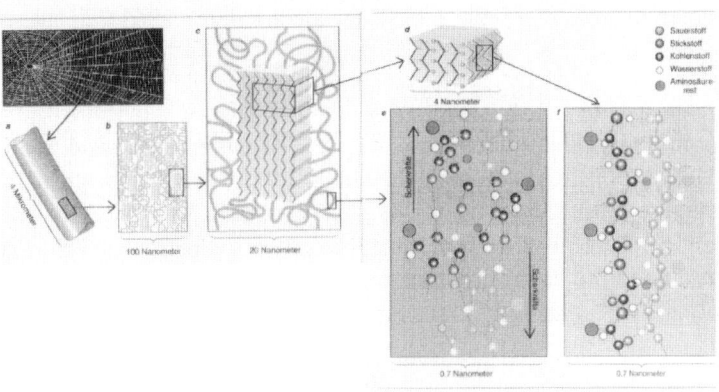

Fig. 51. Molecular structure of spider silk[111]

Of interest is the extremely limited number of substances utilized (carbon, oxygen, hydrogen, nitrogen, and amino acid remnants). Just as fascinating is the disposal of no longer needed threads, which are re-consumed by the spider as protein-containing nourishment. The properties of spider silk far exceed those of all existing technical materials. The silk of some spiders is twice as resistant to tearing as steel and up to fifty times more elastic than nylon. Spider silk has been intensively studied by researchers around the world for some time, including work carried out for the U.S. Army in the hope of developing better bullet-proof vests and improved parachutes. Three completely different approaches to large-scale production of spider silk are presently being pursued. One strategy is the "milking" of test spiders in the laboratory, resulting in up to 200 meters of spider silk per day. The animals are anesthetized with CO_2 and immobilized. Ethically, this may be (with respect to the spiders) a less than ideal solution, but with a view to possible environmental impacts, however, milking is not entirely unreasonable and to some extent comparable to the traditional harvesting of wool or silk from the silkworm. The two other strategies are a chemical approach to synthesizing spider silk and the implantation of a spider silk–coded genes in cultivated bacteria., It is already possible to manufacture small quantities of spider silk with both approaches. Incidentally, the three approaches outlined also demonstrate clearly that on the basis of "bionic" inspirations from nature very different technical implementation strategies

---

[111] Source: (Vollrath 1992)

(and the resulting assessments thereof) can be pursued. Learning from nature, the bionic point of origin, is a far from adequate guarantee for an "intrinsically safe environmental technology" (see von Gleich 2001).

5.1.1.2 Biomimetic synthesis: nanobionics

The acquirement of the capability of self-organization (and self-replication) was a significant step in the evolutionary transition from inanimate to animate nature.[112] Early forms of self-organization already present in inanimate nature provided the basis for self-organization in animate nature. This included the capability of molecular orientation on the basis of molecular charge. These include, among others, polar molecules such as phospholipids, with a hydrophilic and a hydrophobic (usually lipophilic) pole. The phospholipids, as key components of cell membranes, are of decisive evolutionary significance. When such molecules are applied, if possible in a single layer, to a water surface, they orient themselves automatically and form an aligned so-called self-assembled monolayer (SAM).[113] In the presence of aligned structures (templates), lipids and lipid-like substances can roll themselves into three-dimensional shapes such as spherical micellae and vesicles or cylindrical tubulin.[114] In Fig. 52, the result of such a three-dimensional self-organizational process is depicted. The micellae consist of a spherical double layer of polar molecules. These supramolecular cages can then be used as starting forms for the production of well-defined nanoparticles, which, for example, can then be utilized as a base material for high-performance ceramics. Thus it may be possible to radically reduce the usually high energy and material expenditures in the manufacture of such particles.

---

[112] Cf., for example, Eigen's studies on the "hypercycle" (Eigen & Winkler 1978) as well as the studies of Maturana and Varela on autopoiesis (Maturana & Varela 1987).

[113] SAMs can be very reliably produced using Langmuir-Blodgett technology.

[114] See VDI-Technologiezentrum 2002 as well as Laval et al. 1995; Singh et al. 1995.

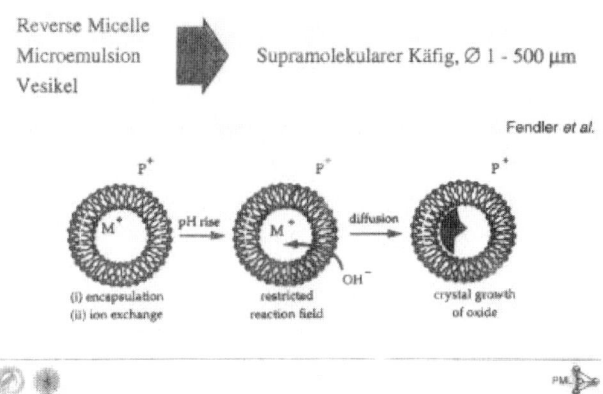

Fig. 52. Synthesis of nanoparticles based on host-guest chemistry[115] (Principle of precipitation of a ceramic layer from solution onto functionalized self-assembled monolayers (SAMs))

Biomimetic materials synthesis clearly refers to using the capability of self-organization and not yet self-replication.

The vast majority of the presently foreseeable uses of the self-organization principles of nanotechnology (nanobionics) are not based on deeply invasive intervention into central control mechanisms (such as cerebral, hormonal, and genetic controls). They are, rather, based on decentralized context control (material gradients) and the application of widely distributed chemical-physical properties of molecules in prepared environments. As long as this is the case, uncontrollable self-reproduction and multiplication will probably be rather unlikely.

5.1.2 Road maps as information and communication concepts for the design process

As already mentioned, the initiative Responsible Care was developed by the German chemical industry. Corresponding initiatives would be well-suited for confronting some of the problems in dealing with hazards in the area of nanotechnology. In the USA, as part of the initiative Chemical Vision 2020 (Vision 2020 2002), stronger industry involvement has already

---

[115] Source: Lecture transparencies Aldinger 1998; see also (Niesen & Aldinger 2001)

occurred and the various challenges arising with the development of nanotechnology in the chemical industry have been identified and labeled. The basis for this is a road map developed for the chemical industry leading up to 2020 . While the actual content of the specific road map is of lesser interest it is important to note the broad spectrum of players from research, regulatory authorities, and various industrial representatives drawn into its preparation. Inclusion of a wide range of actors led to an interdisciplinary discussion of the various dimensions of nanotechnology: safety and environmental aspects were analyzed, but also metrological problems and further research needs were discussed. An analysis of nanotechnology was thus initiated, which, fed by a wide range of research and knowledge sources, identified research needs and calls for action at all levels. In addition to the process of preparing the road map as such, the aspect of dissemination is particularly of interest. This road map including its formulated needs for action was widely distributed by various publications, communicating the findings and problem definitions from the process to the largest possible group of players. This also lends itself to the processes discussed earlier, the development and explication of Leitbild concepts.

The Chemical Vision road map was one of the first activities that demonstrated the opportunities offered by nanotechnology but also described the risks. The acknowledgement that nanotechnology entails not only promises but likewise risks has led to several new endeavors seeking further in-depth analysis of hazards as well as risks and to a significant increase in communication and dialogue activities.

5.1.3 Constructive technology assessment and real-time technology assessment as information and communication concepts for design processes

Constructive Technology Assessment (CTA) (Rip et al. 1995) and real-time TA (Guston & Sarewitz 2001) are highly challenging processes seeking to influence technological development in order to, among other reasons, identify potential adverse effects and to counteract or avoid them by means of suitable measures. Characteristic of these processes is that: i) the relevant players are identified and involved; ii) relevant feedback is made possible in the process; iii) the potential for shaping development is identified; and iv) this potential is exploited.

Given the numerous national as well as international arenas in which nanotechnology development and the dynamics of innovation take place, the assignment of technology assessment is a challenging task from the very beginning. In our view, it is important that these forms of supervision and generation of knowledge (also regarding unintended consequences) be developed. These should take place, as is the case with CTA, real-time TA, and also in the prospective TA approaches applied in this study, concurrently with development.

The design development process should be comprehensive and extend throughout the entire value chain and at the same time consider the temporal dynamics of the product cycle. Thus connections will be established that reach from the research side directly to the disposal phase (including corresponding emissions) and possible changes in the direction of development can be captured in a temporal context. Newer, ecopolitical, approaches, such as the Integrated Product Policy (IPP), are based upon this life cycle assessment approach and attempt to reduce environmental impacts along the entire value chain, for example, by promoting cooperation among the players throughout the product life cycle.

The design process should also take into account the already mentioned aspects of the Leitbild. For example, within the scope of the CTA it is at least in principle possible to work at the Leitbild level – or the implicit orientation level, which is present in every case – and to review "sustainability impacts." In this respect, methods are available to ensure by means of iterative steps greater certainty in the search processes.

The various discourse arenas used for the corresponding processes are a key element. We assume that the design development process is affected by a large number of players with different opportunities for influence, but their influence takes place in multiple arenas. These arenas are of a national nature but are also closely tied to the international level, such that directional influence takes place in a complex multi-tiered system. This does not mean that possibilities for overarching influence and steering development are ruled out, but rather that one has to take into account the multitude of actors in different arenas as a starting point for any efforts to steer the overall technology development in a certain direction (as, e.g., expressed in a certain Leitbild). Various studies have underlined that frequently regional and also national sector players put topics on the agenda and thus can introduce significant directional impulses.[116] Analyses of these dynamics from an ecopolitical point of view have been carried out in

---

[116] Problematic environmental impacts can, for example, become an issue due to the results of some studies; the scientific community may also become aware of problems that are not yet on the agenda, etc. The examples of developmental dynamics in the environmental field are numerous. At the same time the establishment of positively oriented *Leitbilder* has shown itself to be clearly more difficult.

the course of extensive research projects.[117] Such direction-giving impulses are therefore not necessarily dependant on material inputs or power; sometimes ideas and key concepts can have a considerable impact. The (in this context) long-term Leitbild of nanobionics conveys, for example, a series of perceptions that were developed in the short-term Leitbilder also suggested above.

The role of images is of paramount importance to the development of the overall enthusiasm for nanotechnology (and therefore its promotion and support). The visibility of the nano-world has supported engineering fantasies closely associated with fundamentally positive images of precision, for example, "the end of environmental pollution," "abundance of resources," etc. That the reality frequently looks much different, however, is a well-known phenomenon in the area of technological development. However, it is crucial that the – for the most part inseparable – opportunities and hazards be recognized early on and that as a matter of principle the conditions for the attainment of these goals be made a part of constructive Leitbilder.

The course of public discussion to date on the opportunities offered by and hazards associated with nanotechnology has made evident that concrete measures for dealing with potential risks (apart from the reactions to what are viewed by many as possible long-term developments in the form of "foresight guidelines") have only just now been addressed, at a relatively late stage. Likewise, research efforts have yielded until now presumably only a limited understanding of these potential risks. The temporal negligence leads to the problem that negative side effects of technological development in general, and of nanotechnology in particular, usually become apparent only after substantial developments have taken place; a significant time lag between established technologies and knowledge about possible detrimental effects exists on the research side.

However, problems associated with nanotechnologies are now being more systematically investigated. A systematic research program with respect to health risks is developing out of initial warnings, assessments, and knowledge transfer from epidemiological investigations.[118] At the same time, it must be noted that the level of current knowledge is limited and will remain so for the foreseeable time. "Quantitative toxicity studies on

---

[117] Beise, M. et al. (2004) were able, for example, to demonstrate the accruement and diffusion of environmentally friendly technologies based on, among others, environmental regulations and their dissemination.

[118] A summary of the current state of knowledge can be found in Oberdörster et al. (2005a); for the systematization of the problems from a toxicological viewpoint cf. Oberdörfer (2005b: 65).

engineered nanomaterials are still relatively sparse" (Oberdörster et al. 2005b: 7).

Non-governmental and hybrid players

The attention paid in the USA relatively early on to hazards associated with nanotechnologies (research at Rice University) shows that even selective examinations in light of experience with other publicly controversial technologies can already exert significant influence. A more or less systematic investigation into environmental effects introduces and brings topics to the public agenda that otherwise might have been ignored. The creation of a corresponding "institution" is therefore an essential aspect of an integrated approach.

The activities of international NGOs such as the ETC Group underline the importance of the non-governmental players and their ability to set agendas. Irrespective of an assessment of the specific activities of this NGO, it was able to demonstrate that the possible "dark side" of a technology had not been sufficiently worked out and, in particular, that communication with the public (other than with regards to the positive aspects) had not been sought out. At the very least, the media assertiveness of such players suggests gaps in research as well as in communication. The literature study by Howard, commissioned by ETC Group, increased the suspicion that researches, firms, and regulators were blocking out (or had given insufficient attention to) potential risks. In that respect, the study was of considerable importance as it scientifically supported the misgivings of ETC Group by a recognized professional.

The International Council on Nanotechnology (ICON) is based at Rice University and financed by a handful of large corporations, including the global reinsurer Swiss Re. ICON's activities focus in part on providing an Internet platform, organizing some events, and releasing statements on topics relevant to nanotechnology, for example, on the nanotechnology politics of the EU.

The International Risk Governance Council (Geneva),[119] which is concerned with risks of all kinds (particularly so-called systemic risks) has made nanotechnology one of its current priorities. The board members are from government ministries and corporate associations; the participants are from business, various government agencies, and last but not least, an American NGO. It provides a forum for discussion and attempts to offer an

---

[119] IRGC is a public-private partnership in which governments, industry, and academia can freely discuss such issues and together design and propose appropriate risk governance recommendations that have relevance to both developed and developing countries. (http://www.irgc.org/irgc/about_irgc/)

overview of the national nanotechnology politics and to develop recommendations for action specifically with regard to potential risks. The financial support comes above all from business (no nanotechnology firms) and several nations have also contributed.

The Woodrow Wilson Center in the USA has likewise become an essential player in the nanotechnology field. Funded by an endowment, the center, under the direction of David Rejeski, from early on arranged events and conducted analyses, which proved to be path-breaking with respect to side effects in the developing field of nanotechnology. Among other aspects, the center's output showed that the regulatory system could not be characterized as adequate with respect to nanotechnology and that an imbalance existed between the intensive public promotion of nanotechnology and the necessary accompanying research into the potential risks. The Woodrow Wilson Center also distributes information such as an overview of nanotechnological research and nanotechnology-based products. Their information activities, for example, provide regulatory authorities with information on the corresponding products. It is obvious that such institutions do contribute to the availability of the basic information for any risk assessment.

The institutions mentioned represent only a small fraction of the growing nanotechnology scene that has developed in recent years. The brief reference to these organizations is intended to demonstrate that although the "traditional" players continue to be important, increasingly a hybrid field of players is developing that is leading and significantly influencing the public discourse on nanotechnology. The traditional understanding of the distribution of tasks between states and private actors is changing obviously, and in this context national borders play a more and more limited roll. Instead, differentiated fields of discourse are to be seen, whose development dynamic can be appropriately characterized by the term "nanogovernance" (see Petschow/Weizsäcker/Rosenau 2005).

These hybrid organizations are usually significantly more dynamic in taking up new topics and consequently better able to influence the discourse. Nonetheless, these players cannot assume the functions of government. Against the backdrop of the international dimension of technology development, regulatory regimes solely at the state level are almost no longer conceivable. Organizations such as the OECD and the ISO standards bodies are also increasingly being given a larger role to play. However, these have only recently set up committees to deal with nanotechnology – or here, rather, nanoparticles (see below).

The manner in which discourse about an emerging technology is carried out is a further important aspect in dealing with new technologies (particularly when it is a line of technology that eventually will become ubiquitous). Looking at the British example – the deployment of a task force on nanotechnology by the Royal Society – two aspects appear to be significant. First, a relatively broad participatory process was adopted, and both scientists and the general public were included. Secondly, by involving researchers from various fields, differing states of knowledge could be made accessible, thus overcoming the often one-dimensional view of technological opportunities and possibilities; possible adverse effects could therefore be recognized and dealt with from early on.

Such open processes, representing a forum of diverse opinions and viewpoints, can increase the attention given to safety, health, and environmental aspects as well as sustainability-oriented approaches.

The need for such discourse is also increased by the information dynamic achieved. Discussions of nanotechnologies are for the most part globally intertwined. Today, a vast number of information platforms facilitate dissemination of information with unprecedented speed. This applies not only to NGOs but also to all other private and public organizations pushing the discourse on nanotechnologies. One example of this information dynamic, which is also of considerable interest to business, is the 31 March 2006 report from the Federal Institute for Risk Assessment (BfR) in Germany. Several poisoning incidents had been reported in connection with the product Magic Nano, a glass/ceramic sealing spray intended for use in the bathroom and purported to contain nanoparticles. This product was consequently withdrawn from the market. The cause of these poisoning incidents has as of this writing not been fully resolved; authorities have not yet ruled out the possibility that nanoparticles in the product might be responsible for the problems reported.[120] Regardless of the actual outcome in these incidents, the dynamic of the information dissemination process turns out to be extremely relevant. What in German newspapers was only a marginal note (seven lines in the Berliner Zeitung of 31 March 2006, for example) merited in the Washington Post (6 April 2006) an article of more than 40 lines, including queries of various nano-experts. The report was likewise discussed and evaluated on the various Web pages of several organizations (for example ICON, 1 April 2006).

In summary, the level of sensitivity with respect to the existence of potential or alleged nano-risks is clear, as is the significance of the international exchange of information. At the same time, it is evident that although the presumed positive effects associated with "nano" were being used to advertise the product, it is neither clear to what extent it contained nanoparticles – the producer claims that this is not the case (see Small

---

[120] The producer of "Magic Nano" has meanwhile indicated that the product contains no nanoparticles (cf. Small Times 2006).

Times 2006) – nor to what extent the poisoning symptoms could be traced back to the alleged nanoparticles. The case furthermore demonstrates that the term "nano" apparently can be used arbitrarily.

5.1.4 Upstream communication: the integration of citizens/consumers in technological development

The so-called "upstream communication" approach, developed in Great Britain, attempts to involve the ordinary citizen and consumer in the communication process in order to early on address and sound out the substantial obstacles facing newer technologies (for example "green genetic engineering") that have arisen in Great Britain. The upstream communications process is based on the idea that the public acceptance of new technologies is often poorly developed[121] and that new technologies ultimately require public acceptance. The concept "upstream" is perhaps a poor choice of words, as it could give the impression that essentially a one-way communication stream is intended. In principle, and the approaches in Great Britain support this view, the crucial aspect of creating discourse is integrating members of the community and consumers into the technological development process. Specifically, in Great Britain a citizens' jury (the NanoJury) was established; it developed its own assessments in dealing with nanotechnology (here predominantly nanoparticles). This lay assessment was intended to identify from the viewpoint of the average citizen what the future of nanotechnologies should be. The NanoJury called for the labeling of nano-products, so that a freedom of choice would exist, and furthermore the support of risk analysis research. These initiatives offer the possibility of achieving a new level of public engagement and feedback, which can assist in providing the regulatory authorities as well as business a certain degree of certainty with respect to direction.[122]

---

[121] Regardless of acceptance, the knowledge of new technologies is also poorly developed. Macoubrie (2005), for example, points out that the public's knowledge and awareness of nanotechnology is almost non-existent. But at the same time, it is clear that there is no rejection in principle of new technologies (based on "nano"). Appraisals instead tend to be positive; however, safety plays a significant role in this context.

[122] Cf. www.nanojury**Fehler! Textmarke nicht definiert.**.uk; a similar project is being carried out in Germany by the Federal Institute for Risk Assessment (BfR) (UfU/IÖW 2006).

5.2 Player-specific approaches to shaping development

In the standard apportionment into inter-player and player-specific approaches, the latter focuses on the possibilities for action available to the players business and state. On the part of the state, it is a question in this context of possible forms of regulation of business, addressing, above all, the question of risk minimization.

5.2.1 Player business: integration of safety, health, and environmental aspects into comprehensive quality management throughout the value chain

A fundamental approach must target the economic players, the companies themselves. As with the governmental institutions, firms bear a significant responsibility for minimizing hazards and risks and ensuring the risk management of nanotechnological products.

In particular, there is the necessity of integrating the associated safety, health, and environmental aspects into management strategies and systems and into the daily internal processes of doing business. In the course of the ever increasing use of integrated management systems, obtaining the maximum in safety, environmental compatibility, and eco-efficiency should be the goal. Further strategies for minimizing risks in business are, for example, the replacement of dangerous substances by less dangerous alternatives, the utilization of closed, sealed production plants and other technical measures to minimize product and process emissions.

Furthermore, in order for the market players to take on greater responsibility – for example, within the future framework of the European Chemicals Regulation System REACh – the development of new forms of communication and cooperation along the value chain are required. The industrial users of chemical substances and nanoscale systems will be required to take on a much greater responsibility than before and to actively search for low-risk solutions when introducing new materials (the substitution rule). That means that the ascertainment of risks from nanoscale materials within the individual firm, from suppliers and customers (processing and use) and also in the corresponding disposal of products will become a fundamental obligation of each firm (due diligence). This responsibility can only be achieved by close communication among the players (producers, manufacturers, and/or users) within the value chain, because the risks are dependent upon the properties of the materials as well as the corresponding application contexts.

Such a communication pathway, cutting across the European and international markets, in turn necessitates a common standard for the appraisal, assessment, documentation and communication of substance and nano risks. A significant development option lies in the recognition that communication along the value chain (up- and downstream) offers a chance for greater customer-oriented innovation.

5.2.2 Player business: sustainable nanodesign in research and development

The DIN report ISO/TR 14062 Environmental Management – Integrating Environmental Aspects into Product Design and Development (German Institute for Standardization (DIN) 2003) specifies the essential elements of sustainable design. These design principles as a rule must be contextualized and therefore it is crucial to put these principles into specific terms for nanotechnology. Such design possibilities are, of course, dependant on the results of the research; nevertheless there exists, as a rule, a great deal of freedom in these early phases of development.

R&D departments hold the opportunity to influence design; they possess the contextual working knowledge of the processes and procedures and, on the basis of the known positive/negative effects of nanotechnologies (such as particles), can exert decisive influence on potential problems in subsequent stages of development.

In environmental product design, the entire product life cycle becomes an object of the design process. The consequences of alternatives in the stages of raw materials procurement, manufacture, use, and disposal of products are weighed one against the other and compared to other design requirements. Environmental product design implies a greater responsibility on the part of the designers and design engineers. In the design process, right from the start, material and energy inputs, emissions (air, water, waste), minimization of risks and pollutants, as well as product service life and usage, are all looked at and considered over the entire product life cycle. In addition to traditional requirements, goals such as extending useful service life, improving material and energy efficiency, minimizing risks, and improving recyclability are therefore likewise a part of the product development focus.

Starting points for a nano-specific approach to implementing product design criteria can be found, for example, in the VDI (Association of German Engineers) guideline 2243 "Recycling-oriented product development" (VDI 2002) as well as in handbooks and guidelines on environmental design (Geißler et al. 1993; Quella 1998; Umweltbundesamt (UBA) 2000).

Guidelines that focus on the issues of risk analysis and minimization of materials usage should also be addressed. For example, the German Federal Environment Agency (UBA) guidelines on the utilization of environmentally compatible substances (Ahrens et al. 2003) offer risk analysis assistance. By allowing for relevant risk factors, a materials assessment establishing the basis for risk avoidance and minimization strategies becomes possible. Material properties such as persistence, bioaccumulation, aquatic toxicity, chronic vertebrate toxicity, and (environmental) mobility are all relevant to the criteria being assessed; likewise, methods of application and usage, which can be characterized by means of criteria such as quantity, operative conditions of use, and indirect releases.

The idea, however, is not to qualify nanotechnologies as problematic per se but rather to provide assistance with assessment questions in the current situation of extreme uncertainty. And as the necessary research work currently undertaken will not lead to substantial results in the short-term, the uncertainty will not be resolved quickly. However, sensitivity to the problem of uncertainty and the information about possible more or less general guidelines for action (such as the avoidance of particle emissions, the recommendations with respect to measures to safeguard health, and the avoidance of water-soluble substances) can at least provide a certain direction for movement.

In this context, cooperative agreements between business and NGOs could occur and deliberately bring external critical know-how into a company's internal development processes. A case in point is the cooperative arrangement in the USA between DuPont and the Environmental Defense Fund (Environmental Defense 2005). Taking the potential positive benefits of nanotechnologies – particularly with respect to its potential for providing environmental relief – as its starting point, an agreement was reached that looks to these positive benefits but at the same time is intended to catch potential risks as early as possible. The NGO, acting under the motto "Getting things right – the first time," wants to take control of the strained relationship with a development that offers potentially positive environmental benefits as well as risks and, accordingly, address the problems of nanotechnologies and also search for suitable ways to minimize risks from early on.

5.2.3 Player government: federal regulatory approaches

The following observations with regard to the general legal situation and possible requirements for amendment or modification focus primarily on nanoparticles.

The discussions about regulating nanotechnologies also are taking place partly at the meta-level. The question being debated is whether comprehensive regulation is necessary. The viewpoints range from the demand for a moratorium to limit research and applications of nanotechnology (etc group 2002) to arguments suggesting that such measures would actually be counterproductive (Reynolds 2002). And there are debates such as those already put forward about specific design principles as to how potential consequential problems can be avoided. This study is also adressing this latter level of the debate. We assume that – regardless of scientifically established knowledge with respect to effects on the targets – extensive evidence can be generated by means of technology characterization alone, thus justifying risk management measures in a wider sense that could increase acceptance of the precautionary principle[123] without noticeably impeding essential social or technological innovation processes.

Justifications for possible legal regulatory measures result largely from the information that was identified in the course of the investigation of potential adverse effects of nanotechnologies, particularly with respect to possible environmental and health impacts.

A basic issue is whether a new legal framework would be required for the regulation of nanotechnologies (for example, a specific body of nanotechnology law), whether the existing regulatory system is sufficient, or whether the existing system must be adapted to deal with the scope of problems that could arise with the new technology (see, for example, Davies 2005).[124]

We see potential problems in the short-term with respect to the (eco)toxicity of nanoparticles. We generally assume that sufficient regulatory instruments are already available in order to deal with these nanotechnology problems, i.e., specifically the chemical regulations and, with a view to very long-term developments, in some cases also genetic engineer-

---

[123] The precautionary principle is also, of course, subject to numerous interpretations. Basically, the controversy has to do with the question as to the extent to which governmental measures independent of scientific risk analyses are justified. Since it deals centrally with the assessment problem, the interpretation of the precautionary principle proves to be controversial. As a rule, the interpretation of the precautionary principle takes place in each of the corresponding contexts and must therefore be adapted to the specific requirements. (PrecauPri and its strategies). It is a matter of the question of upon whom the burden of proof falls – those wishing to bring new products to the market or those who may potentially be affected (for example, regulatory authorities). The existing differentiations in chemical regulations are based on these distinctions.

[124] The requirements of the various regulatory approaches and areas turn out to be very different. (see Paschen et al. 2003))

ing regulations, such that at the formal level no new regulations are necessary.[125] On the political side, there are differing opinions on the strategically minded thesis that it would be simpler to pass a new body of "nano-law" than to try to adapt the various existing regulations in order to make them "nano-compatible."[126]

In the future, nanotechnology will be extensively deployed as a cross-sectional technology and therefore touch the most varied spheres of regulation, from occupational safety and health to pharmaceutical and cosmetics regulation, environmental protection policies, and more. Thus an extensive need for adaptation of these regulations will arise. The top priority according to those experts consulted is above all the area of occupational safety and health.

The problems that regulatory approaches are intended to address relate primarily to the properties of nanoparticles. A significant aspect of the problem is that already known substances can have fully new properties when manufactured at the size of nanoparticles. At the same time, completely new classes of materials can be produced (e.g., nanotubes). Furthermore, our literature review and our expert survey showed that across-the-board statements about nanoscale structures and substances or classes of substances are not yet possible. In addition to the known problems of substance assessments, the nanoscale realm brings forth an additional dimension of hazards. This has far-ranging consequences, as previous risk assessment methods must be expanded and new measuring methods and test procedures suitable for nanoparticle specifications must first be developed.

Current regulatory practice is still far away from being able to deal adequately with the potential problems of nanotechnologies or, initially and more specifically, nanoparticles; a reliable classification system for recording these structures and substances does not yet even exist. A US study showed clearly (Wardak 2003) that new materials such as nanotubes cannot adequately be categorized using the existing classification schema of the Toxic Substances Control Act (TSCA). Even where classifications were made, the nano substances have so far been assigned to different categories due to a lack of systematic classification rules.

---

[125] This is the view of a series of experts. In particular, the proceedings of the Royal Society made clear that currently the overwhelming majority of players do not believe that a new regulatory framework custom-designed for nanotechnology is necessary.

[126] This thesis was introduced at a hearing on nanotechnology organized by the German political party Bündnis 90/Die Grünen held on 1 March 2004 in Berlin.

A systematic category assignment in the application procedure has apparently not yet been established.[127] Bergenson (2006) reviews ongoing discussions and finds that in the context of the TSCA, a distinction between "nano" and "macro" is in fact not made. When a substance is registered, for example, the chemical substance is defined by its molecular identity; this definition does not address a substance's physical and chemical properties. Bergeson likewise continues: "Under TSCA Section 8(e), any person who manufactures, imports, processes, or distributes a chemical substance or mixture and who obtains information that reasonably supports the conclusions that the chemical substance or mixture poses a substantial risk of injury to human beings or to the environment must notify EPA of such information immediately." Hence, notification regarding nano-substances must at the very least be made when it must be assumed that there are associated hazards involved. However, this does not lead to a systematic reporting of nanoparticles – if no hazard is identified, there is accordingly no current obligation of notification. The first notification to the US EPA of a substance occurred in October 2005, when the EPA "approved the manufacture of a new kind of carbon nanotube under the 'low release and exposure exemption' of the TSCA" (Pellerin 2006).

The situation in Germany (or generally in Europe) is not significantly different. Classification systems derived from chemical regulations that in part also require re-registration in the case of substantial changes in the properties of a substance or material do exist. But in practice this is currently not being done. A discussion with the responsible approval authorities showed that in Germany to date (2004) no substance or material has been re-registered as a nano-material and therefore no corresponding guidelines have been established for the registration of such substances.

The neglect of re-registration points to a second problem: the prerequisites for approval and the reporting requirements are first and foremost geared to substances produced in bulk (by weight!). The appropriateness of this approach is increasingly being questioned (cf. SCENIR 2005: 55). The planned European REACh System does implement new forms of risk assessment, for example, that identify specific properties of a substance as already being a reason for concern. However, it must be noted that the procedures for approving a substance, at least for substances that are to be produced in smaller quantities, will be significantly reduced. With respect to nano-materials, or more specifically nanoparticles, this means that such an assessment from REACH in its present form, due to the generally minimal quantities as measured by weight, would never be focused on

---

[127] Our written expert survey also makes reference to the necessity of developing a suitable classification system.

unless they were classified per se "very persistent and very bioaccumulative (vpvb)."

But apart from chemical regulations, considerable problems arise with respect to nanoparticle emissions. The litigation concerning the licensing of a plant for the production of particles in southern Germany is a current example. The central question in this context was and is (the case is now to be decided by the Federal Administrative Court), whether there are specific dangers associated with nanoparticles that are comparable to conventional problems – including those associated with ultrafine particles (UFPs) – or whether nanoparticles exhibit entirely different properties and effects with respect to environmental and health risks. Against this background of particle pollution, which is caused above all by combustion processes, it is significant whether it is valid to equate the particles resulting from nanoparticle production with those arising from combustion processes, or whether industrially manufactured "custom-made" particles are a reason for particular concern. Although the regulations in general make no distinction between UFPs and nanoparticles, an attempt was made to deal with this by establishing a commensurate margin of safety for approved emissions. In the existing "background pollution" caused by combustion processes, the impact of nanoparticles, as a rule, actually turns out to be negligible.[128]

Discussions emphasizing regulatory aspects have so far not gone very far. They generally run along the following lines: on the one (extreme) side, there is ETC Group's call for a moratorium on the development of nanotechnology, followed by the demands of the Royal Society (2004) that the emission of nanoparticles be avoided, then the Foresight Guidelines, voluntary stipulations (for example, the US EPA), and finally the current regulatory system, which ultimately does not address nanostructures directly and becomes active only after a hazard becomes known. In principle, these more or less simplistic approaches to the regulation of nanoparticles are the result of the existing extreme uncertainty about potential negative impacts of nanoparticles, which currently does not allow for any general statements with regard to possible health and safety risks (SCENIHR 2005).

---

[128] At the same time the general ecopolitical discussion makes clear that ecopolitical measures are increasingly successful in reducing environmental pollution caused by particulates and other "large particles." At the same time, it is also clear that UFP pollution is increasing, and not least because of the limited agglomeration potential of larger particles. This means that at least in ecopolitical discussions, the particle question (and separately from that of industrially produced particles) will play an increasing role.

Initial analyses of the recording of data on nanomaterials refer to registration and classification problems with nanomaterials (particularly nanoparticles); classifications of the substances with respect to the chemical regulations either did not occur or were undertaken only irregularly. In the course of various interviews, including conversations with regulatory authorities, it was revealed that initial discussions of the problem first took place in 2004.

Currently, the regulatory question with respect to nanotechnology has been only scarcely addressed. A scoping study by Chaudry (2005; see also US EPA 2005) aiming to identify gaps in environmental regulations analyzed the existing regulatory framework with respect to nanotechnology (i.e., nanoparticles); the regulatory gaps found were above all due to a general overall lack of information and data on nanoparticles; contributing were regulatory exemptions (tonnage threshold exemptions) as well as the situation already mentioned several times, that potential effects from nanostructures remain unknown. Important approaches to regulating nanotechnologies are in principle already available (particularly chemical regulations). At the same time, as already described, current regulatory frameworks are not suitable (or only to a limited extent) to deal with the potential problems of nanotechnology. A central aspect that must be mentioned in this context is the current lack of a classification system – at least one that takes the specifics of nanotechnology into account. It is therefore not possible for current legislation to address the particular specifications of nanoparticles. Furthermore, potential risks cannot be assessed currently – or only to a limited extent – because it is an open question whether the results of exsiting lab studies are transferable. These studies urge caution at the same time.

Before classification could proceed, one would first of all have to create the conditions necessary for the classification of nano-substances in order to develop suitable testing and measuring procedures and to establish the basic conditions for a risk assessment. As long as the application contexts in many cases remain unknown and exposure assessments are therefore not possible, it seems necessary to fall back on meta-rules, for example, the precautionary principle. Substantial elements in this respect can also be derived from the EU chemical regulatory approach. In doing so, the tonnage standard (1 t/a) could be a problem that in the case of specific concerns must be viewed in perspective.

The international dimension of nanotechnologies and particularly the regulation of nanotechnologies is turning out to be an extremely important issue. The framework of the TBT Agreement passed by the WTO calls for a convergence of national standards with international standards: WTO members are obliged to use the "International Standards and Guides" as the basis for their technical regulations, standards, and conformity assessment systems. The TBT agreement states that "where technical regulations are required and relevant international standards exist or their completion is imminent, Members shall use them, or the relevant parts of them, as a basis for their technical regulations ... except when such international standards or relevant parts would be an ineffective or inappropriate means for the fulfillment of the legitimate objectives pursued, for instance because of fundamental climatic or geographical factors or fundamental technological problems."(UN 2004 11)

Against the background of international exchanges of goods and services, it is more than the international cooperation of all countries that proves to be essential to avoiding trade barriers. Accordingly it is not surprising that international organizations such as the OECD or standardizations organizations such as ISO have taken up the issue of nanotechnologies relatively early on.

At the same time it must be pointed out that the international exchange of knowledge in the areas of scientific research and business, but also with the NGOs, has contributed to an elevation of the discussions on nanotechnologies to the international level. Scientific results (concerning advantages as well as disadvantages) for nanotechnologies are being rapidly distributed, the dissemination of products is international, and NGO activities, too, have finally achieved a global reach.

The activities of the OECD and ISO make it possible to expedite communications processes and coordinated efforts, making it possible to fill in gaps in our knowledge and develop consistent terminology. However, the international approach could also lead to a premature convergence and standardization that could call into question precautionary approaches at the national level should they not be WTO-compatible.

The OECD led the first international-level discussions of the issue at a meeting in Washington D.C. in December 2005. The gathering took place as part of the Chemical Program of the OECD and had as its goal identification of the human health and environmental safety issues of manufactured nanomaterials (short- to long-term) and sought, above all, to address questions of definitions, nomenclature, and characterization; environmental effects; human health effects; and the regulatory framework.

The activities of the OECD are based on the necessity for international coordination – sharing as well as exchange – and converging baseline information when addressing regulatory frameworks, assessment methodologies, and testing schemes. OECD activities in the area of chemical regulation offer a certain role-model function.

The decision was made to found a "working party" that would "organize and prioritize future activities (management and assessment of NM [nanomaterials] for environment, health safety), development of a database (overtaking Woodrow Wilson Center's and ICON), (with BIAC): creation of a foundation dataset by testing representative nanomaterials."

In the course of the Washington Workshops three break-out sessions were conducted to work on definitions, nomenclature, and characterization; environmental and health effects; and regulatory frameworks.

The documented results of the workshops show that a greater need for convergence and development of definitions, nomenclature, and characterization exists, that a great deal of uncertainty with respect to potential environmental and health effects exist, and finally, with respect to regulatory frameworks, that "regulatory regimes may be adequate in most jurisdictions and although further assessment may be required a voluntary approach gathering and developing data and evidence could complement existing regulatory regimes, help identify any gaps and inform future work." All the same, it may well be possible that "in the long run, a new approach may be needed depending on evidence gathered"; and as the first priority: "develop guidance on 'new' versus 'existing' chemicals."

In the course of international discussions on nanotechnology, the standards organizations have joined in. At the European level, the technical committee CEN/TC 352 was established and at the international level, ISO/TC 229. ISO/TC 229 is taking the lead; CEN/TC 352 will cover those standards areas not addressed by ISO. The founding of ISO/TC 229 followed in November 2005 and includes the following working groups

- WG 1: Terminology and Nomenclature.
  Scope: To define and develop uniform terminology and nomenclature in the field of nanotechnologies. It is intended to facilitate communications to ensure common understanding among interested parties
- WG 2 Measurement und Characterization.
  Scope: Standardization of metrology and test methods (including reference materials) which is used to characterize nano-materials and nano-structures from the aspect of physical and chemical properties
- WG 3 Health, Safety, and Environment.
  Scope: To develop standards in the areas of health, safety, and environmental aspects of nanotechnologies including occupational, environmental, and public exposure and monitoring; engineering controls, personal protective equipment, and other measures to assure safety of workers, researchers, and the public; epidemiological and environmental surveillance protocols; human and ecological bio-kinetics and toxicity; disposal, dispersion, and waste treatment of nano-engineered materials; as well as methodologies, data quality, and data analysis for risk assessment.

It is obvious that dynamic discussions are being held at the international level and moreover that an exchange between the various levels of stakeholders is taking place. The OECD sees in the activities of ISO a complementary, technically oriented process that supports its own work.

The discussion of the potential negative effects of nanotechnology will now also be carried out at the international level. In the context of both the OECD and the standards organizations, nanotechnologies have been taken on as a field of action. In particular, the OECD makes explicit reference to the activities of NGOs such as the Woodrow Wilson Center, which has already performed a great deal of preliminary work with respect to the establishment of a knowledge base with its databases (listing, for example, products on the market containing nanoparticles and scientific publications on hazards). It is therefore to be expected that the discussion process will be pushed forward and priorities for dealing with nanotechnologies will be developed. However, it must also be noted that standardization processes, in the context of ISO, for example, could also take place "too soon." It could be possible that standardization in some cases would lead to hasty conclusions in a field long characterized by uncertainty. The key will be to find pathways that will lead to suitable standards, but at the same time make provision for dealing with the unknown and unknowable.

6 Conclusions, the outlook, and need for action

In order to do justice to the complexity of the problem definition, a three-step approach to prospective technology assessment and design of nanotechnology was utilized in the project.

- **1st approach – prospective**
  Assessment of nanotechnology and its effects by means of a characterization of the technology.
- **2nd approach – process-concurrent**
  Evaluation of sustainability effects utilizing life cycle assessment methods (extrapolating) on typical applications in comparison to existing products and processes.
- **3rd approach – formative**
  Leitbilder as "guiding instruments" in technology development, associated processes, actor-specific concepts.

6.1 Results of environmental assessments of selected nanotechnological application contexts

In the course of the above-mentioned research project, specific products and processes were analyzed to determine to what extent the application of nanotechnological contributions could contribute to environmental relief, and to what extent these possibilities can be realized. This emphasizes the dimension of opportunity offered by nanotechnology with a focus on contributions to resource efficiency and reductions in environmental pollution. It can be seen in the majority of the case studies that nanotechnology offers foreseeable eco-efficiency potentials. As an example, the chosen nanotechnological applications in the display field (OLED, CNT-FED) show higher energy efficiencies in the use phase (in some cases by a factor of two for reduced energy usage) as compared to previous solutions. Even greater potentials for efficiency increases by nanotechnological applications could be seen in selected cases in the area of industrial coatings and lacquers, as well as in catalytic applications. In the lighting field, however, nanotechnology-based products will have to undergo further development before they are able to compete with other energy-saving light sources with respect to resource and energy usage.

At the same time it must be noted, there is no guarantee that processes showing promising results in the life cycle assessment will automatically succeed in the marketplace. This is in part an assessment issue – such processes are still in an early phase of development, but the other reason is that these innovations – like all innovations – often must overcoming inertial forces; for example the enormous "sunk costs" in existing equipment may well inhibit the introduction of new nanotechnologically based processes. Such inertia often inhibits the introduction of innovations and represents a significant obstacle. In this area as well, much depends on the practicality of implementing the opportunities for environmental relief offered by nanotechnology.

Feeding the results of the prospective product assessment profile back into the development process makes possible further optimizations in the design of processes and products.

- The concurrent, prospective life cycle assessment has proven itself to be an important instrument in identifying significant elements of sustainable technological development. The availability of life cycle tools and data for relevant manufacturing processes would substantially facilitate the preparation of environmental assessment profiles.
- The environmental advantage is not necessarily economically enforceable; a sustainability-oriented design must also be aware of the obstacles to market entry.

6.2 Technology characterization and context analysis

Innovations, including those oriented toward sustainable development, always come with certain risks. The step "technology characterization" in the study is based on the assumption that technologies are not neutral with respect to the environment, that specific characteristics can give us clues to potential risks, that evidence derived from the technology characterization can serve as the basis for risk management measures following the precautionary principle, and that such evidence can and should be incorporated into development of the formative model.

A basic message of this analysis was that environmentally non-contained processes should be avoided at least until more is known about the behavior of the relevant substances and their interaction with the "tar-

gets." This approach corresponds, for example, to the handling of many substances within the framework of the chemical industry regulations.
- Non-contained applications involving nanoparticles are to be avoided.
- Critical aspects include, for example, the mobility of nanostructural materials, permeability of cell membranes, solubility in water and fat, biopersistence, and accumulation.

Much more difficult is the next step in the selected approach, the consideration of the application contexts and the analysis and characterization of affected systems. Technological impacts are revealed not only by the "character of a technology" but, rather more so, by the corresponding application contexts (in both quality and quantity). A prospective assessment of the potential environmental and health impacts of nanotechnology therefore must not only concentrate on the technology characterization and nano-specific impacts, it must also consider social trends and the specific application contexts.

At this point in time, the risks of nanotechnology are difficult to evaluate. In the current debates it is above all the long-term visions that are dominating discussions with respect to risks and opportunities. The current opportunities for nanotechnology, to the extent that they are significant to production, exist essentially in the use of the potential of nanoparticles and, above all, their unique properties. The unique properties of nanoparticles and nanostructures should, on the one hand, be developed for technical, economic, and environmental reasons, but they bring with them, on the other hand, potential risks. Within the scope of a comprehensive risk assessment, it is not possible to quantify these risks when exposure data is lacking. Research data on fine and ultra-fine particulate from a hazards analysis of toxicological and epidemiological aspects is available, however the data, in many cases, are not – and that should be emphasized – based on industrially manufactured nanoparticles, but rather particulate from combustion processes. Investigations of the behavior of nanoparticles in the human body indicate that airborne nanoparticles can penetrate the body and brain through the lungs and even the bulbus olfactorius. Little is known so far about the possible effects. Toxicological investigations have demonstrated adverse effects in laboratory experiments; however the results of these experiments are still somewhat contradictory.

With respect to the environmental impacts of nanotechnology, current knowledge is even less developed; there has been particularly little research done on the behavior of the particles, for example, in the various environmental compartments. Central questions arise particularly with a view to the behavior of nanoparticles in the air, for example the question to what extent and in what time frame do agglomerations take place and how stable are they? More rapidly formed and more stable agglomerations would in some cases "defuse" potential problems.

The analysis of various production processes makes clear that for many of them the issue of emissions in the air pathway is less problematic. Likewise, over the entire life cycle of nanotechnological products, only a limited number of problems – at least from the present viewpoint – are to be seen. At the very least, in the case of nanoparticles incorporated into a solid matrix, particle emissions are hardly to be expected. All in all, much too little is known about these issues; there is an enormous need for further research.

In the majority of applications investigated so far, the results presented indicate that particulate emissions, when handled appropriately throughout the life cycle, very likely do not represent a fundamentally unsolvable problem. In manufacturing particles, the problems seem to be manageable, at least in principle. During the use phase, the particles should essentially be immotile; at the end of the life cycle – dependent, of course, on the recycling method – there should not be any emissions; however further research is necessary. This is fundamentally the approach that was established in the framework for the analysis (avoidance of non-contained applications).

In the assessment of nanotechnology with respect to opportunities and risks, it became clear in the expert survey that the opportunities far overshadow the risks. However, we are also reminded that the necessary concurrent investigations and risk assessments must be carried out.

The need for research includes:
- Toxicological and ecotoxicological analyses within the scope of integrated research programs
- Investigation of the behavior of nanoparticles and nanostructural surfaces in the environment and their classification

Despite the problem of comparing different risk structures, with the help of the technology characterization carried out here and in accordance with current knowledge with respect to foreseeable nano-specific effects (including self-organization), no "especially great cause for concern" could be established. The foreseeable and anticipatable risks appear most likely to be comparable to the risks associated with (synthetic) chemistry; these are, accordingly, by no means negligible, as chemical risks have historically shown themselves to be quite substantial. Early technological assessment and design intervention as well as early regulation and precautionary measures to avoid many or most of the mistakes made in the field of chemistry are therefore advisable. The REACh system, being developed by the EU as a part of ongoing chemical regulatory reform, may well anticipate the needed steps in the risk analysis (with the exception of the necessary transition from a regulatory approach based on bulk weight to one based on particle quantity). In addition, much can be learned about risk management from the chemical industry and chemical handling procedures. However one must be cautious with this analogy that weaknesses in current risk management procedures are not also carried over, as risk management in the chemistry industry exhibits enormous loopholes, for example in the implementation of the precautionary principle (keyword: "intrinsically safe substances," techniques, and application systems; keyword: "substitution principle").

Above and beyond the case studies investigated, which very much focused on inorganic applications, there is a particular need for further research on:
- the application of self-organization in the inorganic as well as organic fields with a view to still-to-be-done and much-further-reaching eco-efficiency potentials.
- "active nanosystems" and, furthermore, on the ongoing, long-term application fields of self-organization, particularly in the case of a possible merger of nanotechnology and bio- and/or genetic engineering, with a view to the question of a gradual transition from self-organization to self-reproduction.

This fundamental assessment of nanotechnology likewise leaves unresolved questions concerning risk assessment and management with respect to nanotechnology. This concerns, first of all, the question of classification of substances; this currently is being done unsystematically or not at all. Here, as the expert survey also demonstrated, there is a need for action. Furthermore, it must be determined how the unique properties of the nanoparticle can be addressed through modifications to the existing regulatory system (beginning with the systems of measurement) and how these properties can appropriately be dealt with. Here, too, the chemical regulatory framework offers significant clues, although in its current form it does not address the specifications of the nanoparticle.
- The existing regulations are as a rule applicable to nanotechnology, but in their current form, they do not address the nanoparticle. A classification system for nanotechnological products must be established; appropriate methods of measurement must be developed.
- An adaptation of REACh is possible, but at the same time a series of further accommodations in the areas of environmental and health law must follow; the issue of worker safety must be given priority.

6.3 Process- and formative model–based design and development

Technological development is not controllable by political intervention – or, if so, only in a very limited way; instead, the interaction of all possible players usually results in a path of development that can be constructively influenced by, among other things, the use of a formative model as a control instrument. Since nanotechnology is still for the most part in an early phase of development, there still exists, in principle at least, a large degree of freedom with which to steer research efforts in a "sustainable" direction.

The formative model is not a simple instrument of "intervention." The explication and further elaboration of such models is to a large extent a communication process taking place among the greatest possible number of players. The three proposals for sustainable nanotechnology outlined in this project, each having a different scope of action, could serve as material for the communications process: "Resource-efficient Nanotechnology," "Consistent and Intrinsically Safe Nanotechnology," as well as the model "Nanobionics," with its long-term orientation. The initiation of processes leading to the development and shaping of formative models through the requisite procedural steps is desirable and sensible. This can happen in the course of various methods, for example in the course of a constructive TA.

Furthermore, processes, such as those most recently initiated and being carried out in England by the Royal Society (compiling knowledge and conducting information and communications work), can also serve as an example for other countries.

The science-technology and business players could – much like the "Chemical Vision" developed in the course of EU debates and the "road maps" developed in the US – agree on an integrated concept for nanotechnology development that certainly, with broad participation, would sound out and bring to bear the opportunities of this line of technology and at the same time identify the technological and economic, but also environmental and health, risks.

If widely held expectations for nanotechnology (and the innovations to be derived from it) are to be realized with respect to making significant contributions toward a sustainable economy, i.e. if the intention to accordingly guide and shape the respective innovation processes is to be taken seriously, then further ongoing, concurrent assessments of technology, process, and product development are essential, first of all, with respect to precaution-oriented risk management, secondly, with respect to the intended full realization of the sustainability potential associated with this line of technology.

At the operative level, in addition to working on and with a formative model, the known assessment, information, and communication instruments seem to be essentially well-suited for generating guidelines for action, particularly for small and mid-sized businesses (guidelines, development directives, etc.).

- The early phase of the technology development offers in theory a great deal of potential for steering development in the direction of sustainability – this potential should be realized. The development of formative models offers itself as a valid approach.
- Concurrent and design-oriented processes must be implemented (for example by means of CTA or real-time TA). Open communications processes, in which science, business, and non-governmental organizations are included, are to be recommended. Roadmaps for nanotechnological development are an appropriate instrument for integration and providing direction. Concurrent prospective life cycle assessments should also be utilized and their results should be fed back into the developmental process.
- As players in these processes, businesses have a fundamental responsibility, which – particularly in more recent EU environmental policy approaches such as REACh or in Integrated Product Policy (IPP) – has been repeatedly emphasized. The approaches mentioned assume at least a shared responsibility and in that respect the industry must clearly fulfill its responsibilities. Specifically, integrated, comprehensive management concepts, in which the aspects of health, safety, and environment (HSE) are recognized as an element of quality assurance, need to be developed. For the support specifically of small and medium-sized businesses, guidelines for sustainability-oriented design of products and processes would also be helpful.

Appendix

Table 41. Survey of selected nanotechnology-based products and application status[129]

---

[129] Source: (Paschen et al. 2003)

224 Appendix

Nanotechnology-based Product	Nanosubstance and/or Structure	Manufacturing Method	Industrial Sector	
SURFACE TREATMENT AND FINISHING				
mechanical properties				
tribological layers	hard coatings	ultra-thin oxide or hard coatings	gas phase deposition	machine-building, vehicle construction
	nano-composite hard coatings	nanocrystallites and nanoparticles in amorphous matrix	gas phase deposition	
	superhard C coatings	ultra-thin coatings; nanocrystallites in amorphous matrix	PECVD: vacuum arc	
hard disk C topcoating	corrosion protection and tribology	ultra-thin coatings; nanocrystallites in amorphous matrix	PECVD: vacuum arc	IT
scratch-resistant coatings on soft materials	Oxide coatings on paper, polymers, wood, textiles	ultra-thin coatings, nanopores	sol-gel process	packing, construction, paper industries
scratch-resistant coating	e.g., glass and metal	inorganic-organic hybrid polymers	sol-gel process	vehicle construction, textiles
anti-adhesive coatings	antifriction coatings, anti-graffiti coatings	ultra-thin polymer coatings	self-organization	consumer goods, construction, ship-building
impact-resistant window glass	composite layer between glass panes	ultra-thin polymer coatings		automotive engineering, construction
polishing compounds	for chem.-mech. polishing of silicone wafers	nanostructured oxides (e.g., SiO_2, CeO)	flame hydrolysis	IT

Application	Function	Nanomaterial	Process	Industry
thermal and chemical protective coatings thermomechanical protective coatings	turbines, engines	nanodispersion/nanocomposites; ultra-thin coatings	gas phase deposition sol-gel process	air and space industry, vehicle construction
protective diffusion coatings for computer chips	hinders the inward diffusion of Cu in Si	ultra-thin titanium nitride coatings		IT
gas barrier coatings	PET drink containers, food packaging	ultra-thin oxide coatings; inorganic-organic hybrid polymers	gas phase deposition; sol-gel process	food industry
	food industry packaging, tennis balls	nanoparticles (silicates); nanocomposites (SiO2 in polymer)	mineral extraction; flame hydrolysis	food industry, consumer goods
wettability superhydrophobic protective coatings	Diamor®, SICON®, simultaneously wear-resistant	ultra-thin coatings, nanocrystallites in amorphous matrix	vacuum arc, gas-phase deposition	div.
self-cleaning textiles	hydrophobic and oleophobic	ultra-thin hybrid-polymer coatings	sol-gel process	textiles
inkjet printer paper	hydrophilic	amorphous nanoparticles (oxides)	flame hydrolysis	IT
Biocidal coatings antibacterial/biocidal protective coatings	antibacterial	ultra-thin coatings	soft lithography	food industry; construction, medicine, ship-building
	controlled release of biocides	oxide nanopores	sol-gel process	
	photocatalystic production of hydroxyl radicals	nano particle, ultra-thin oxide coatings	sol-gel process, possibly CVD	

Application	Property/Effect	Nanomaterial	Process	Industry sector
antibacterial textiles	antiseptic coatings	polymer nanocoatings	e.g., soft lithography	textiles, medicine
	controlled release of biocides	oxide nanopores	sol-gel process	
optical properties anti-reflective coatings for glass	solar cells, storefronts, etc.	e.g., nanoporous SiO2 coatings	sol-gel process	construction, energy
switchable transparency coatings	window panes	ultra-thin nanocrystalline coatings	div.	construction, automotive engineering
paper whiteners	colorings, color effects	nanoparticle (oxides) as coating carbon black nanoparticles	flame hydrolysis flame soot process	consumer goods textiles
lacquers/industrial coatings, adhesives Paint and lacquer coatings	color brilliance, hue, color effects, etc.	nano particle and ultra-thin oxide or carbon coatings	flame soot process, wet-chem. deposition, colloid chemistry	automotive engineering, consumer goods
preservative coatings for wood	UV resistance	TiO2 nanocrystals	flame hydrolysis	construction
adhesives	optimal flow characteristics	amorphous nanoparticles (oxides)	flame hydrolysis	automotive engineering, consumer goods
	silver conductive adhesive	metal nanoparticles	div.	automotive engineering, IT

CATALYSIS, CHEMISTRY, MATERIALS SYNTHESIS

catalyzers	nanoparticle catalyzers	nanocrystalline metals or semiconductors; cluster	precipitation, sol-gel process...	chemicals, automotive engineering, environmental engineering
	nanoparticle photocatalyzers	nanocrystals, cluster (oxide or metal)	precipitation, sol-gel process...	environmental engineering, chemicals
	zeolites as catalyzers	nanoporous hydrated silicates (zeolites)	self-organization	chemicals, environmental engineering
	molecularly imprinted polymers (MIPs)	nanoporous polymers and oxides	molecular imprinting	chemicals, medicine, pharmaceuticals, environmental engineering
	nanoporous carriers	nanoporous zeolites, aerogels, etc.	slip casting, sol-gel process, extrusion...	chemicals, automotive engineering, environmental engineering
	nanoreactors	organic molecules, polymers, oxides	self-organization...	chemicals
	biocatalysts	nanopor. oxides, ultra-thin coatings	sol-gel process	biotech, environment engineering
molecular sieves	water treatment, etc.	sintered nanoparticle materials	self-organization, colloid chemistry	chemicals, environmental engineering, food industry
ultrafilter membranes	water treatment, etc.	nanoporous polymers	self-organization	
chromatography materials	column materials	nanoporous oxides	sol-gel process	environmental engineering, chemicals, medicine/pharmaceuticals
	MIPs	nanoporous polymers	molecular imprinting	
ion exchangers		nanoporous zeolites	self-organization precipitation,	pharmaceuticals, chemicals
antifoaming agents		silicic acids	flame hydrolysis	

227

CONVERSION AND UTILIZATION OF ENERGY

Application	Material	Process	Field
fuel cells	nano-ceramic solid-material electrolytic membranes for improved ionic conductivity	Sintered-nanoparticle ceramics	div.	power engineering, automotive engineering
super condensation	nanoporous electrodes	nanotubes, nanoporous carbon aerogels		power engineering
superconductors	CeO_2 nanocoatings in YBCO cable	ultra-thin oxide coatings	e.g., gas phase deposition	power engineering, medicine
ceramic furnace igniters		nanoceramic (SiC)	div. + injection molding + pressureless sintering	power engineering
glow plugs	diesel engines	Si nanoparticles in amorphous ceramic	polymer pyrolysis	automotive engineering
nano-tips for field-emission switches	ignition systems	piezoceramics (nanoscale dia. tips)		automotive engineering
lubricants	MoS_2 fullerenes additives in oils	inorganic fullerenes oxide nanoparticles (CeO_2)	self-organization div.	automotive engineering automotive engineering

CONSTRUCTION

Application	Material	Process	Field
building materials	nanoscale concrete aggregates	nanoparticles (SiO_2)	from filter dust, flame pyrolysis	construction engineering
	polymer aggregates for cement	nanoparticles	not known	construction engineering
facings, fittings, housings	nanoparticle-strengthened thermoplastics	nanoparticles	e.g., furnace process	automotive engineering, IT, consumer goods
tires (particle)	carbon black, silicates	nanocrystals and nanoparticles (carbon black, SiO_2)	flame-assisted deposition	automotive engineering

Application	Product	Nanostructure	Process	Field
components for medical devices	particle-reinforced elastomers	amorphous nanoparticles (oxides)	flame hydrolysis	medicine
components for rocket engines	high-temperature resistant particle-reinforced elastomers	nanoparticle precious metals and oxides	red. of AgCl / hydrolysis of silicon alkoxide	air and space travel
additional functionality of structural materials				
antistatic textiles	upholstery fabrics, dress fabrics	aggregate, nanostructure carbon particles	flame-assisted deposition	automotive engineering, textiles
tinted glass		metal nanoparticles	sol-gel, ion implantation	construction engineering, consumer goods
INFORMATION PROCESSING AND TRANSMISSION				
nanoelectronic components	(C)MOS, bipolar transistor high electron-mobility transistor diffusion barrier layer between Cu and Si metallic nanoconductors	ultra-thin lateral nanostructure from semiconductor ultra-thin lateral nanostructure from compound semiconductor ultra-thin coating of nanocrystalline nitrides metallic nano-wiring on surfaces	PVD, CVD, lithography PVD, CVD, lithography (ammonia) oxidation, sputtering, lithography and electrochemical deposition	IT
displays	organic LEDs / OLEDs	ultra-thin coatings from organic compounds	PVD, spin coating, immersion	IT, vehicle construction
semiconductor lasers	quantum-film laser	ultra-thin semiconductor coatings	CVD, PVD	IT
coil cores for mobile communication and GPS receivers	partially crystallized metallic glasses	nanocomposite of metal-nanocrystallites in amorphous metal matrix	partial nanocrystallization	IT
color photocopiers, color laser printers	drum	nanocomposite of nanoparticle pigments in conductive polymer		IT

NANOSENSORS AND ACTUATORS

sensors				
	GMR magnetic-field sensors	ultra-thin coatings of metal	CVD/PVD/MBE	automotive engineering, mechanical engineering
AFM and STM tips for analytics and nanostructuring		nanoscale tips of carbon nanotubes	gas-phase deposition + molecular manipulation, etching	nanotechnology, research
magnetic-field sensors	GMR sensors	ultra-thin coatings of metal	CVD/PVD/MBE	IT
	spin-polarized scanning tunneling microscope	ultra-thin coatings and ultrafine tips of metals or semiconductors	PVD, etching	analytics
	magnetic force microscope	ultra-thin coatings and ultrafine tips of metals or semiconductors	PVD, etching	analytics
	near-field optical microscope	ultra-thin coatings and ultrafine tips of metals or glass	PVD, etching	analytics
infrared sensors	quantum-well infrared photodetector	ultra-thin coatings	CVD, PVD, self-organization	IT, armaments
temperature sensors	scanning thermal microscope	ultrafine metallic tips		IT, analytics
chemically sensitive scanning probes	chemically sensitive STM and AFM tips	ultra-thin coatings, ultrafine tips, and molecules	PVD, etching, molecular manipulation	analytics
electrochemical scanning probes	scanning tunneling microscope	ultrafine metallic tips	etching process	analytics
	atomic force microscope	ultra-thin coatings; ultrafine tips	etching process	analytics

lab-on-a-chip systems, bio-(gen) chips	SAMs for cell immobilization Semiconductor quantum dots as (fluorescence) markers	molecules, ultra-thin coatings nanocrystals, ultra-thin coatings (metals, semiconductors, polymers)	self-organization, soft lithography div.	pharmaceuticals, medicine, armaments
shock absorbers, vibration dampers	ferrofluids	nanocrystals of oxides and metals	div.	automotive engineering, air and space travel, armaments
LIFE SCIENCES				
antifoaming agents	silicic acids		precipitation, flame hydrolysis	pharmaceuticals, chemicals
nanoparticle markers	quantum dots	metallic nanocrystals	div.	pharmaceuticals, medicine
active ingredient carrier	liposomes	organic molecules (lipids)	self-organization	pharmaceuticals, medicine
	porous oxide particles	nanoporous oxides (Al2O3)	aluminum anodizing	pharmaceuticals, medicine
gene transfer system for genetic engineering	gene gun	DNA-coated nanoparticles of metal	div.	medicine/pharmaceuticals, biotechnology
biological decontamination agents	reactive nanoparticles	oxides (e.g., MgO) loaded with halogens (e.g., Cl2Br2)	not known	pharmaceuticals, medicine, armaments
cosmetics articles	effects pigments	ultra-thin nanoparticle coatings (TiO2, Fe2O3, SiO2...) nanoparticles (SiO2)	wet-chemical deposition colloid chemistry	cosmetics
	dermal sensation wrinkle lighteners	amorphous nanoparticles (oxides)	wet-chemical deposition	
suntan cream	TiO2 nanocrystals	nanocrystalline TiO2	flame hydrolysis	cosmetics

Table 42. World extraction of chromium ore[130]

Chromium ore: annual global extraction from 1987 to 1996 in 1,000 t/a										
Year	1987	1988	1989	1990	1991	1992	1993	1994	1995	1996
Europe	5648.9	5704.2	6394.7	5966.2	5527.9	5438.4	4201.3	4528.1	4887.5	3691
Africa	4552.2	4987.5	5772.1	5326.6	5820.6	4056.8	3246.5	4274.6	5941.9	5779.2
America	243.4	282.1	320.8	307.4	433	499	357.6	458.6	462.4	458.5
Asia	926.8	1065.8	1385.7	1243.5	1376.5	1392.4	1357.8	1611.8	2333.4	2032.5
Oceanica	61.8	70.3	60.3	6.2		8.2				
World total	11,433.1	12,109.9	13,933.6	12,849.9	13,158.0	11,394.8	9,163.2	10,873.1	13,625.2	11,961.2

[130] Source: (Eggert et al. 2000)

Table 43. Basis data for 1 kg application-ready industrial coating[131]

	1 K CC	2 K CC	Water CC	Powder CC	Nanocoat
Primary energy input [MJ]	87.00	97.70	62.90	124.00	146.55
Energy resources [MJ]					
Crude oil	48.75	49.05	31.08	63.46	73.58
Natural gas	29.93	38.44	24.52	38.53	57.66
Uranium ore	3.82	4.20	3.05	9.12	6.30
Coal	2.95	4.00	2.82	8.57	6.00
Brown coal	1.25	1.65	1.12	3.41	2.48
Hydro-electric	0.16	0.19	0.13	0.38	0.28
Other	0.19	0.20	0.18	0.37	0.29
Material resources [kg]					
Rock salt	0.217	0.045	0.124	0.379	0.067
Limestone	0.014	0.005	0.062	0.192	0.008
Air emissions					
CO_2 [kg]	3.164	3.828	2.685	4.953	5.742
NM VOC [g]	11.960	15.400	8.831	12.810	23.100
Methan [g]	7.221	7.753	5.440	11.410	11.630
NO_x [g]	6.492	8.145	5.048	8.395	12.218
SO_2 [g]	4.325	6.785	3.841	8.163	10.178
CO [mg]	636	1036	595	990	1554
Particulate [mg]	458	1646	399	812	2469
HCl [mg]	41	41	47	144	62
HF [mg]	10	8	11	33	11
Water emissions					
Water discharge [l]	99.690	81.160	81.790	160.200	121.740
CSB [g]	6.738	10.560	6.608	8.144	15.840
TOC	4.122	6.047	3.781	5.187	9.071

---

[131] Source: Harsch and Schuckert 1996 and authors

BSB [g]	1.649	1.244	1.281	1.421	1.866
Solids [g]	0.876	6.620	0.661	1.971	9.930
HC [mg]	103.500	99.850	62.020	140.100	149.775
NaCl [g]	271.900	91.850	71.290	172.500	137.775
Iron [mg]	12.760	12.330	8.247	18.290	18.495
Nickel [mg]	1.232	1.155	1.193	3.353	1.733
Chromium [mg]	0.236	0.225	0.150	0.326	0.338
Lead [mg]	0.209	0.616	0.168	0.491	0.924
Copper [mg]	0.125	0.117	0.097	0.229	0.176
Cadmium [mg]	0.036	0.035	0.028	0.066	0.052
Waste Industrial waste [g]	45.17	42.71	37.59	165.90	64.07
Household waste [g]	16.29	13.59	13.17	17.54	20.39
Hazardous waste [g]	33.60	29.90	55.17	87.77	44.85
Radioactive waste [g]	1.12	1.27	0.83	2.30	1.90

Table 44. Data on application[132]

Primary energy input [MJ/body]	1 K CC	2 K CC	Water CC	Powder CC	Nanocoat
Spray booth	206.20	207.70	164.50	160.80	164.50
ZW-Trockner	117.50	117.50	187.60	117.50	117.50
Dryer	428.40	428.40	545.80	428.40	428.40
Peripherals	8.76	8.76	8.76	8.76	8.76
Coating treatment	51.80	51.80	51.80	41.46	51.80
Water discharge	18.53	18.53	48.57	-	18.53
Exhaust air	50.87	50.87	-	-	-

	1 K CC	2 K CC	Water CC	Powder CC	Nanocoat
Resource use [MJ/body]					
Crude oil	16.88	16.91	17.50	8.17	14.83
Natural gas	320.00	320.10	298.60	293.10	280.78
Uranium ore	212.00	212.80	226.10	167.30	186.66
Coal	123.10	123.50	131.30	97.13	108.33
Brown coal	83.13	83.42	88.65	65.60	73.17
Hydro-electric	7.46	7.50	7.96	5.89	6.58
Resource use [kg/body]					
Rock salt	0.000	0.000	0.000	0.000	0.000
Limestone	0.433	0.434	0.461	0.342	0.381
Air emissions [kg/body]					
CO_2	42.10	41.60	40.00	33.00	36.49
NM VOC	0.029	0.022	0.02	0.015	0.019
Methane	0.0844	0.0846	0.0867	0.0692	0.074
NO_x	0.0694	0.0653	0.0515	0.0392	0.057
SO_2	0.0288	0.0289	0.0299	0.023	0.025
CO	0.0117	0.0112	0.0098	0.0074	0.010
Particulate	0.01022	0.00996	0.00953	0.00689	0.009
HCl	0.00146	0.00146	0.00156	0.00115	0.001
HF	0.000525	0.000527	0.00056	0.000414	0.000

Water emissions [g/body]

---

[132] Source: Harsch and Schuckert 1996 and authors

Discharge water [l/body]	3709	3718	3809	3047	3261
CSB	3.774	3.78	3.875	1.2	3.316
TOC	8.527	8.531	8.109	6.887	7.483
BSB	0.5609	0.5618	0.5774	1.953	0.493
Solids	7.563	7.589	8.063	5.971	6.657
HC	0.02228	0.02229	0.02252	0.002837	0.020
Chloride + NaCl	33.92	3.396	3.444	1.483	2.979
Iron	0.07883	0.07909	0.08394	0.6074	0.069
Nickel	0.001333	0.001337	0.001415	0.0009721	0.001
Chromium	0.000871	0.0007321	0.0007762	0.0005519	0.001
Lead	0.00607	0.006091	0.006473	0.00479	0.005
Copper	0.0005642	0.0004243	0.00045	0.0003218	0.00037
Cadmium	0.00006807	0.00006829	0.00007229	0.00005013	0.00006
Waste [g/body]					
Industrial waste	43.23	43.38	45.23	34.47	38.05
Household waste	3.64	3.64	3.70	0.70	3.19
Hazardous waste	1.25	1.26	1.34	0.99	1.10
Radioactive waste [mg]	1.67	1.68	1.78	1.32	1.47

The dehydrogenation of ethylbenzene to produce styrene is a reversible endothermic balance reaction:

$C_6H_5C_2H_5 \leftrightarrow C_6H_5C_2H_3 + H_2$ $\Delta H^{600°C} = 124.9$ kJ/mol (1)

In addition to this main reaction, the following important secondary reactions (at a minimum) also take place (Lieb & Hildebrand 1982; Schoen 2002):
The deaklylation of ethyl benzene to benzene (1%):

$C_6H_5C_2H_5 \rightarrow C_6H_6 + C_2H_4$ $\Delta H^{600°C} = 101.8$ kJ/mol (2)

The formation of toluene (2%):
$C_6H_5C_2H_5 + H_2 \rightarrow C_6H_5CH_3 + CH_4$ $\Delta H^{600°C} = -64.5$ kJ/mol (3)

Generation of carbon by cracking:
$C_6H_5C_2H_5 \rightarrow 8\,C + 5\,H_2$ $\Delta H^{600°C} = 1.72$ kJ/mol (4)

And as the reaction occurs in the presence of water:
The water vapor molecules react with ethyl and methane, yielding carbon monoxide and hydrogen:

$2\,H_2O + C_2H_4 \rightarrow 2\,CO + 4\,H_2$ $\Delta H^{600°C} = 232.1$ kJ/mol (5)
$H_2O + CH_4 \rightarrow CO + 3\,H_2$ $\Delta H^{600°C} = 224.6$ kJ/mol (6)

The carbon monoxide conversion:
$H_2O + CO \rightarrow CO_2 + H_2$ $\Delta H^{600°C} = -36.0$ kJ/mol (7)

The water vapor fulfills the important task of gasifying the carbon accumulating in the catalyzers through equation (4), thus preventing the blocking of the active centers of the catalyzer that would lead to its gradual deactivation.

$C + 2\,H_2O \leftrightarrow CO_2 + 2\,H_2$ $\Delta H^{600°C} = 99.6$ kJ/mol (8)

Fig. 53. Secondary reactions of styrene synthesis

Table 45. Gross raw materials and fuel demand in mg for the production of 1 kg styrene[133]

Fuel type	Input in mg
Crude oil	660,000
Gas/condensate	1,100,000
Coal	57,000
Metallurgical coal	350
Lignite	5,800
Peat	3
Wood	1
Biomass	720

Table 46. Gross demand for source materials in mg for the production of 1 kg styrene[134]

Raw material	Input im mg
Air	73,000
Barytes	<1
Bauxite	790
Bentonite	230
Calcium sulphate	23
Chalk	<1
Chromium	<1
Clay	16
Dolomite	11
Feldspar	<1
Ferromanganese	1
Fluorspar	7
Granite	<1
Gravel	3
Iron	940
Lead	1
Limestone	2,100
Nickel	<1
Nitrogen	28,000
Olivine	8
Oxygen	27
Phosphate as P2O5	<1
Potassium chloride	1
Rutile	<1
Sand	110

[133] Source: Boustead 1999
[134] Source: Boustead 1999

Shale	64
Sodium chloride	1,800
Sulphur (bonded)	30
Sulphur (elemental)	87
Zinc	<1

Table 47. Gross water demand in mg for the production of 1 kg styrene[135]

Source	Use for processing (mg)	Use for cooling (mg)	Totals (mg)
Public supply	3,600,000	-	3,600,000
River canal	3,400	59,000	62,000
Sea	440,000	125,000,000	126,000,000
Unspecified	140,000	28,000,000	28,000,000
Well	16	840	850
Totals	4,100,00	154,000,000	158,000,000

Table 48. Gross emissions for the production of 1 kg styrene[136]

Emission	From fuel production (mg)	From fuel use (mg)	From transport operations (mg)	From process operations (mg)	From biomass use (mg)	Totals (mg)
Dust	690	380	6	32	-	1,100
CO	160	1,300	66	25	-	1,600
CO_2	340,000	2,000,000	19,000	4,300	-670	2,400,000
SO_X	1,900	4,600	260	60	-	6,800
NO_X	3,300	5,700	140	89	-	9,200
N_2O	<1	<1	-	-	-	<1
Hydrocarbons	490	310	41	1,500	-	2,400
Methane	5,800	1,700	-	360	-	7,900
H_2S	-	-	-	<1	-	<1
HCl	16	<1	-	<1	-	16
Cl_2	-	-	-	<1	-	<1
HF	1	<1	-	<1	-	1
Lead	-	<1	-	<1	-	<1
Metals	1	2	-	<1	-	3
F_2	-	-	-	<1	-	<1
Mercaptans	-	<1	-	<1	-	<1
Organo-Cl	-	-	-	<1	-	<1
Aromatic-HC	-	-	-	140	-	140
Polyclic-	-	-	-	<1	-	<1

---

[135] Source: Boustead 1999
[136] Source: Boustead 1999

HC Other organics	-	-	-	3	-	3
CFC/HCFC	-	-	-	1	-	1
Aldehydes	-	-	-	<1	-	<1
HCN	-	-	-	<1	-	<1
H2SO4	-	-	-	<1	-	<1
Hydrogen	-	-	-	46	-	46
Mercury	-	-	-	<1	-	<1
Ammonia	-	-	-	<1	-	<1
CS2	-	-	-	<1	-	<1
DCE	-	-	-	<1	-	<1
VCM	-	-	-	<1	-	<1

Table 49. Gross solid waste resulting from the production of 1 kg styrene[137]

Type	From fuel production (mg)	From fuel use (mg)	From process operations (mg)	Totals (mg)
Mineral	11,000	-	3,500	15,000
Mixed industrial	330	-	780	1,100
Slags/ash	1,400	60	530	1,900
Inert chemical	<1	-	1,100	1,100
Unspezified	16	-	440	460
Construstion	<1	-	13	13
Metals	-	-	18	18
To incinerator	-	-	6	6
To recycling	-	-	750	750
Paper & board	-	-	92	92
Plastics	-	-	2	2
Wood waste	-	-	<1	<1

Table 50. Gross water pollutants in mg resulting from the production of 1 kg styrene[138]

Emission	From fuel production (mg)	From fuel use (mg)	From transport operations (mg)	From process operations (mg)	Totals (mg)
COD	4	-	-	300	310
BOD	3	-	-	18	21
Acid	<1	-	-	39	39

---

[137] Source: Boustead 1999
[138] Source: Boustead 1999

Dissolved solids	47	-	-	42	89
Hydrocarbons	30	<1	-	46	76
NH4	<1	-	-	7	8
Suspenden solids	16	-	-	300	320
Phenol	3	-	-	<1	3
Al+++	-	-	-	100	100
Ca++	-	-	-	1	1
Cu++	-	-	-	<1	<1
Fe++/Fe+++	-	-	-	<1	<1
Hg	-	-	-	<1	<1
Pb	-	-	-	<1	<1
Mg++	-	-	-	<1	<1
Na+	-	-	-	700	700
K+	-	-	-	<1	<1
Ni++	-	-	-	<1	<1
Zn++	-	-	-	<1	<1
Metals unspezi-fied	<1	-	-	450	450
NO3-	-	-	-	3	3
Other nitrogen	<1	-	-	7	7
CrO3	-	-	-	<1	<1
Cl-	-	-	-	3,500	3,500
CN-	-	-	-	<1	<1
F-	-	-	-	<1	<1
SO4-	-	-	-	270	270
CO3-	-	-	-	200	200
Phosphate as P2O5	-	-	-	<1	<1
Arsenic	-	-	-	<1	<1
DCE	-	-	-	<1	<1
VCM	-	-	-	<1	<1
Detergent/oil	-	-	-	53	53
Dissolved Cl2	-	-	-	<1	<1
Organo-chlorine	-	-	-	<1	<1
Dissolved organics	-	-	-	30	30
Other organics	-	-	-	1	1
Sulphur/sulphide	-	-	-	1	1

Table 51. Road Map of Display Technology

	R&D Theme	2000	2005	2100
Enlargement	Size	32	50	50
Energy Saving	Power Consumption	1	1/3	1/5
High Resolution	Deg. Of Minuteness	15~40 ppi	40~50 ppi	50~100 ppi
Luminescence Efficiency	LCD	2 lm/W	3 lm/W	4 lm/W
	PDP	1.2 lm/W	5 lm/W	10 lm/W
	CNT-FED	NA	7 lm/W	14 lm/W
	Organic EL	1~2 lm/W	7 lm/W	10 lm/W
	CRT	2 lm/W	2 lm/W	2 lm/W
Power Consumption	LCD	140 W	120 W	100 W
	PDP	300 W	200 W	100 W
	CNT-FED	70 W	70 W	40 W
	Organic EL	NA	60 W	30 W
	CRT	200 W	230 W (HVTV)	200 W (HVTV)
Stability of Pixels	LCD	2%	2%	2%
	PDP	2%	2%	1%
	CNT-FED	15%	2%	1%
	Organic EL	2%	2%	1%
	CRT	1%	1%	1%
Lifetime	LCD	60,000 h	60,000 h	60,000 h
	PDP	8,000 h	10,000 h	10,000 h
	CNT-FED	10,000 h	10,000 h	10,000 h
	Organic EL	100 h	6,000 h	10,000 h
	CRT	10,000 h	10,000 h	10,000 h
Speed of Pesponse	LCD	15 msec	13 msec	10 msec
	PDP	10 msec	10 msec	10 msec
	CNT-FED	2~3 msec	2~3 msec	2~3 msec
	Organic EL	1 μsec	1 μsec	1 μsec
	CRT	2~3 msec	2~3 msec	2~3 msec
Resolution	LCD	40 ppi	50 ppi	70 ppi
	PDP	30 ppi	40 ppi	50 ppi
	CNT-FED	30 ppi	50 ppi	60 ppi
	Organic EL	75 ppi	130 ppi	200 ppi
	CRT	30 ppi	60 ppi	60 ppi

Table 52. Breakdown by mass of the main components of a 17″ CRT monitor[139]

Material/Component	Sub-component	Mass (kg)	
Lead oxide glass		9.76	
	Lead		0.45
Steel		5.16	
Plastics		3.04	
	Polycarbonate (PC)		0.92
	Styrene-butadiene co-polymer		0.83
	Polyethylene ether (PEE)		0.74
	Acrylonitrile-butadiene-styrene (ABS)		0.32
	High-impact polystyrene (HIPS)		0.15
	Triphenyl phosphate		0.05
	Tricresyl phosphate		0.02
	Phosphate ester		0.01
Printed wiring boards (PWB) and components		0.85	
Cables/wires		0.45	
Aluminum (heat sink)		0.27	
Nickel alloy (invar)		0.27	
CRT shield assembly		0.24	
Ferrite		0.17	
Deflection yoke assembly		0.15	
Demagnetic coil		0.13	
Video cable assembly		0.11	
Power cord assembly		0.11	
Electron gun		0.10	
CRT magnet assembly		0.08	
Audio cable assembly		0.07	
Frit		0.07	
Solder		0.03	
Phosphors		0.02	
Aquadag		0.02	
Other (misc.)		0.06	
TOTAL		21.16	

---

[139] Source: Socolof et al. (2001)

Table 53. Breakdown by mass of the main components of a 15″ LCD monitor[140]

Material/Component	Subcomponent	Mass (kg)	
Steel		2.53	
Plastics		1.78	
	Polycarbonate (PC)		0.52
	Poly(methyl methacrylate) (PMMA)		0.45
	Styrene-butadiene copolymer		0.36
	Polyethylene ether (PEE)		0.3
	Triphenyl phosphate		0.09
	Polyethylene terephthalate (PET)		0.06
Glass		0.59	
Printed wiring boards (PWB) and components		0.37	
Cables/wires		0.23	
Aluminum (heat sink, transistor)		0.13	
Solder (60% tin, 40% lead)		0.04	
Color filter pigment		0.04	
Polyvinyl alcohol (PVA) (for polarizer)		0.01	
Liquid crystals, for 15 LCD unspecified 2		0.0023	
Backlight lamp (cold cathode fluorescent lamp, CCFL)		0.0019	
	Mercury		3.99E-06
Transistor metals, other (e.g., Mo, Ti, MoW)		0.0019	
Indium tin oxide (ITO) (electrode)		0.0005	
Polyimide alignment layer		0.0005	
Other (e.g., adhesives, spacers, misc.)		0.0031	
TOTAL		5.73	

[140] Source: Socolof et al. (2001)

Table 54. CRT inventory of the various life-cycle phases per functional unit[141]

Inventory type	Up-stream	Manu-factoring	Use	End Of Life	Total	Per Unit
Inputs						
Primarymaterials	15.80	421.00	219.00	-3.32	653.00	kg
Ancillary materials	2.11	3.54	3.47	10.70	19.80	kg
Water	554.00	11400.00	1140.00	-27.30	13100.00	kg
Fuels	8.00	428.00	0.00	-2.95	433.00	kg
Electricity	73.20	129.00	2290.00	0.23	2490.00	MJ
Total energy	366.00	18300.00	2290.00	-128.00	20800.00	MJ
Outputs						
Air pollutants	30.00	183.00	449.00	2.47	664.00	kg
Wastewater	17.00	1510.00	0.00	-3.65	1520.00	kg
Water pollutants	0.81	20.10	0.07	-0.06	20.90	kg
Hazardous waste	489.00	113.00	0.00	8.28	9.46	kg
Solid waste	9.55	81.20	83.30	-1.66	172.00	kg
Radioactive waste	0.00	0.00	0.00	0.00	0.00	kg
Radioactivity	38 E06	3,780,000	48 E06	4,800	89,8 E06	Bq

Table 55. LCD inventory of the various life-cycle phases per functional unit[142]

Inventory type	Up-stream	Manu-factoring	Use	End Of Life	Total	Per Unit
Inputs						
Primary materials	235.00	49.20	80.10	-2.19	362.00	kg
Ancillary materials	1.06	204.00	1.29	2.11	208.00	kg
Water	263.00	2150.00	425.00	-18.00	2820.00	kg
Fuels	14.70	25.80	0.00	-1.95	38.60	kg
Electricity	34.60	316.00	853.00	0.16	1200.00	MJ
Total energy	633.00	1440.00	853.00	-84.40	2840.00	MJ
Outputs						
Air pollutants	112.00	64.80	168.00	1.30	346.00	kg
Wastewater	8.57	3120.00	0.00	-2.41	3130.00	kg
Water pollutants	0.46	1.23	0.03	-0.04	1.68	kg
Hazardous waste	0.01	4.64	0.00	1.64	6.29	kg
Solid waste	13.10	12.60	31.10	-4.42	52.30	kg
Radioactive waste	22.10	3140.00	31.10	-5.23	3190.00	kg
Radioactivity	12E06	10,2E06	17,9E06	3,400	40,1E06	Bq

---

[141] Source: Socolof et al. (2001)
[142] Source: Socolof et al. (2001)

Table 56. Emission factors for the generation of 1 kWh of electrical energy (Energiemix Deutschland)[143]

Emissions	Amount	
Carbon dioxide	638.948	g
Methane	1.587	g
Sulfur dioxide	686.148	mg
Nitric oxide	583.023	mg
Carbon monoxide	222.744	mg
Dust / particulate	76.660	mg
NMVOC	39.016	mg
Nitrous oxide	24.316	mg
Hydrogen chloride	19.836	mg
Hydrogen fluoride	1.237	mg
Nickel	0.038	mg
Lead	0.036	mg
Mercury	0.012	mg
Arsenic	0.011	mg
Cadmium	0.003	mg

---

[143] Source: GEMIS 4.1 (2003)

Table 57. Flow of material for per reference light quantity (RLQ)[144]

Material	GL	EL	EL Longlife	LED today	LED future
Sand [g]	178.96	40.70	22.79		
Soda [g]	71.02	8.73	4.89		
Crude oil [g]	45.79	190.56	106.71		
Silica sand [g]				9.25	2.55
Dolomite [g]	35.22	1.03	0.58	0.001	0.00
Feldspar [g]	28.28	4.70	2.63		
Barite [mg]				22.27	6.13
Waste rock [g]				593.97	163.61
Rock salt [g]		0.16	0.09	4.352	1.20
Potash [g]	3.47	9.04	5.06		
Glass cullet [g]	87.49	15.68	8.78		
Paper [g]		0.41	0.23		
Bleached wood pulp [g]	12.52	6.35	3.56		
Waste paper [g]	90.57	46.77	26.19		
Scrap metal [g]		2.36	1.32	0.338	0.09
Oxygen [g]		22.17	12.41		
Oxygen [dm^3]	21.990	22.75	12.74		
Nitrogen [mg]			0.00	0.003	0.00
Oxygen [dm^3]		14.95	8.37		
Hydrogen [g]	0.08	1.03	0.58		
Water [kg]				54.43	14.99
Air [kg]				2.03	0.56
Soil [mg]				37.24	10.26
Calcium carbonate [g]		1.23	0.69	0.670	0.18
Barium sulfate [g]	3.80	0.11	0.06		
Cerium concentrate [g]	0.21	0.33	0.18		
Magnesium carbonate [g]				1.00	0.28
Ammonium nitrate [mg]				0.14	0.04
Barium carbonate [g]		4.25	2.38		
Barium titanate [g]		0.20	0.11		
Bauxite ore [g]	57.00	11.00	6.16		
Precious-metal ore			0.00	1851.84	510.09

[144] Source: Mani (1994), Gabi 4 database (2001) and authors' own calculations

Iron ore [g]		11.96	6.70	2.979	0.82
Nickel ore [g]				207.08	57.04
Lead – zinc ore [mg]				53.52	14.74
Copper ore [mg]				0.33	0.09
Zinc – copper ore [mg]				26.12	7.19
Zinc – lead – copper ore [mg]				5.67	1.56
Zinc – lead ore [mg]				0.11	0.03
Yttrium [g]		2.29	1.28		
Europium [g]		0.12	0.07		
Cerium [g]		0.09	0.05		
Terbium [g]		0.06	0.03		
Antimony [g]		0.05	0.03		
Lead oxide [g]	5.43	0.27	0.15		
Borax [g]		2.07	1.16		
Aluminum fluoride [g]	0.21	0.04	0.02		
Mg/AlO4 [g]		0.63	0.35		
NaOH [g]		0.02	0.01		
Caustic soda [g]	5.10	0.99	0.55		
Sulfur [mg]				1.08E-03	0.00
Iron [mg]				1.73E-05	0.00
Bentonite [mg]				6.90	1.90
Tungsten [g]	0.13	0.02	0.01		
Brass [g]	1.28	4.61	2.58		
Lead [g]	1.81	0.34	0.19		
Tin [g]	0.39	0.52	0.29		
Molybdenum [g]	0.03				
Nickel [g]	0.55	0.34	0.19		
Copper [g]	0.55	11.20	6.27		
Mercury [g]		0.01	0.01		
Silicon [g]		0.33	0.18		
Solvent [g]		0.18	0.10		
Hydrochloric acid [g]		1.21	0.68		
Sodium salt [g]		0.06	0.03		
Energy UCPTE 88	1.51	20.65	11.56		
Ceramic [g]		0.52	0.29		
Argon [g]	0.001				

Chlorine – rubber [g]		0.20	0.11
Kaolin [g]	0.72	0.37	0.21
Coating color [g]	6.37	3.23	1.81
Rolling oil [g]	0.12	0.10	0.06
Anodes [g]	5.11	0.99	0.55
Lacquer [g]		0.14	0.08
Bromine [g]		0.39	0.22
Electrolyte [g]		1.02	0.57
Fuse [g]		0.15	0.09
Diac [g]		0.10	0.06
Ancillary materials, additives [g]	3.81	2.02	1.13
Additional materials [g]		1.11	0.62
Other [g]	0.357	0.86	0.48

Table 58. Air emissions for per reference light quantity (RLQ)[145]

	GL	EL	EL Longlife	LED today	LED future
Carbon dioxide [g]	352,148	78,242	70,523	233,673	64,660
Dust / particulate [g]	45.938	10.481	9.074	28.555	7.901
Carbon monoxide [g]	123.061	28.656	25.358	81.831	22.643
Hydrocarbons (incl. polycyclic aromatic hydrocarbons) [g]	4.537	8.222	4.604	0.000	0.000
Nitric oxide [g]	322.878	74.238	65.944	214.590	59.377
Nitrous oxide [g]	13.618	3.268	2.847	8.894	2.461
Sulfur dioxide [g]	382.636	91.968	80.183	256.384	70.937
Aldehyde [mg]	1.200	6.400	3.584	0.001	0.000
Other organic compounds (incl. VOC, NVOC, halogen-containing) [g]	21.462	5.381	4.644	15.182	4.200
Other inorganic emissions [g]	0.000	0.000	0.000	764.323	210.534

---

[145] Source: Mani (1994), Gabi 4 database (2001) and authors' own calculations

Ammonia [mg]	0.300	2.000	1.120	2.085	0.574
Hydrogen chloride [mg]	10,923.190	2,385.596	2,165.070	7,280.712	2,014.611
Hydrogen fluoride [mg]	683.356	149.041	135.170	483.039	133.623
Lead [mg]	23.184	8.886	6.493	13.367	3.699
Cl2 [mg]	0.003	0.001	0.000	0.000	0.000
Mercury [mg]	6.651	1.711	1.464	4.441	1.229
Fluorine / fluorides [mg]	2.900	0.600	0.336	0.026	0.007
Tar [g]	0.001	0.000	0.000	0.000	0.000
Chlorine / chlorides [mg]	2.400	0.500	0.280	0.000	0.000
Mercaptans [mg]	0.400	0.200	0.112	0.000	0.000
Cadmium [mg]	2.243	2.882	1.743	1.132	0.313
Zinc [mg]	2.720	9.590	5.370	0.607	0.167
Selenium [mg]	1.700	13.020	7.291	0.104	0.029
Arsenic [mg]	8.229	13.060	7.791	4.301	1.190
Nickel [mg]	23.050	20.874	13.279	13.947	3.859
Copper [mg]	5.270	40.350	22.596	0.025	0.007
Iron [mg]	6.350	48.650	27.244	0.336	0.092
Molybdenum [mg]	0.003	0.000	0.000	0.001	0.000
Var. heavy metals [mg]	0.000	0.000	0.000	1.324	0.365
Var. emissions [g]	0.000	0.000	0.000	1.708.521	470.615
Methane [g]	873.070	190.488	173.027	579.401	160.327

References

Abrams BL, Thomes WJ, Bang JS et al. (2003) Degradation of nanoparticulate-coated and uncoated sulfide-based cathodoluminescent phosphors. Rev. Adv. Mat. Sci. 5:139 - 164

Ahrens A, Böhm E, Heitmann K et al. (2003) Leitfaden zur Anwendung umweltverträglicher Stoffe.

Ahrens A, Gleich Av (2002) Von der Kreislaufwirtschaft zur Nachhaltigen Chemie - Leitbilder in der Chemikalienentwicklung und Stoffpolitik.
http://www.subchem.de/startgerman.html, Date accessed: 03/19/2004

Aixtron AG (1999) Navigator. Ausgabe 1, Mai 1999, Aachen

Ali TA, Ghosh AP, Howard WE (1999) Digest 1999. San Jose, SID Int. Symp. (Society for Information Display)

Amaratunga G (2003) Watching the Nanotube: Tiny tubes of carbon could oust plasma in large flat-panel displays. Spectrum online.
http://www.spectrum.ieee.org/WEBONLY/publicfeature/sep03/nano.html, Date accessed: 10/01/2003

Anders G (1965) Die Antiquiertheit des Menschen. Über die Seele im Zeitalter der zweiten industriellen Revolution. Beck, München

Anger G (1982) Chrom-Verbindungen. In: (Hrsg., 1982): Ullmanns Encyklopädie der technischen Chemie. Wiley-VCH, Weinheim

Ankele K, Steinfeldt M (2002): Ökobilanzen. In: Deutscher Wirtschaftsdienst (ed) Der Umweltschutzberater. Loseblattsammlung, Köln

APME (1997) Eco-Profiles of the European Plastics Industry - Report 4: Polystyrene (Second Edition). A Report for The European Center for Plastics in the Environment. APME: Brussels 1997

APME (1999) Eco-Profiles: Styrene. The Association of Plastic Manufacturers in Europe.

Bachmann G (1998) Innovationsschub aus dem Nanokosmos. Technologieanalyse. VDI Technologiezentrum GmbH, Reihe Zukünftige Technologien, Bd. 28, Düsseldorf

Baird D, Bueno O, Lynn S, Ray R, Robinson C (2003) Visual Images in NanoScience/Technology. USC NanoCenter Symposium V, Columbia, SC, 11/21/2003

Basler & Hofmann AG (2002) Nanotechnologie und Life Sciences. Zürich

Baughman RH, Zakhidov AA, de Heer WA (2002) Carbon Nanotubes - the Route Toward Applications. Science. 297:787 - 792

Bea FX, Haas J (2001) Strategisches Management. Grundwissen der Ökonomik. Lucius und Lucius, Stuttgart

Beise M, Blazecjzak J, Edler D, Haum R, Jacob K, Jänicke M, Loew T, Rennings K, Petschow U (2005) Lead Markets for Environmental Innovations. Springer, Berlin Heidelberg New York

Bergeson LL (2006) Reflections on TSCA and Engineered Nanoscale Substances. Contribution to Global Chemical Regulations Conference. Baltimore Marriott Waterfront Baltimore, Maryland, 03/28/2006

Bermudez E, Mangum JB, Wong BA et al. (2004) Pulmonary Responses of Mice, Rats, and Hamsters to Subchronic Inhalation of Ultrafine Titanium Dioxide Particles. Toxicological Sciences. 77:347-357

Bertram D, Weller H (2002) Zwischen Molekül und Festkörper. Physik Journal. 2 / 2002:47-52

Bijker W, Hughes TP, Pinch T (eds.) (1987) The Social Construction of Technological Systems. New Directions in the Sociology and History of Technology, MIT Press, Cambridge

BINE Informationsdienst (2002) Antireflexglas für solare Anwendungen. Bonn

Blankenbach K (1999) Multimedia-Displays - von der Physik zur Technik. Physikalische Blätter. 55 (5):33-38

Blättel-Mink B, Renn O (eds) (1997) Zwischen Akteur und System. Die Organisation von Innovation. Westdeutscher Verlag, Opladen

Blättel-Mink B (2001) Wirtschaft und Umweltschutz - Grenzen der Integration von Ökonomie und Ökologie. Campus, Frankfurt/M.

Bloch E (1973) Das Prinzip Hoffnung. Frankfurt/M.

Blochwitz J (2001) Organic light-emitting diodes with doped charge transport layers. TU Dresden

BLU (2002) Umweltrelevante Inhaltsstoffe in Elektro- und Elektronik-Altgeräten: Bayrisches Landesamt für Umweltschutz.
http://www.bayern.de/lfu/abfall/index.html, Date accessed: 12/02/2003

Borm PJA (2002) Particle Toxicology: From Coal Mining to Nanotechnology. Inhalation Toxicology. 14:311-324

Böschen S (2000) Risikogenese - Prozesse gesellschaftlicher Gefahrenwahrnehmung: FCKW, DDT, Dioxin und Ökologische Chemie. Leske und Budrich, Opladen

Böschen S (2002) Risikogenese: Metamorphosen von Wissen und Nicht-Wissen. Soziale Welt. 53:67-86

Brock T, Groteklaes M, Mischke P (1998) Lehrbuch der Lacktechnologie. Vincentz, Hannover

Brumfiel G (2003) Nanotechnology: A little knowledge... Nature. 424 (6946):246-248

Burden AP (2001) Materials for Field Emission Displays. International Material Reviews. 46:213-231

CEFIC (2006) Capacity and production data for petrochemicals in W. Europe, 2001-2005.
http://www.petrochemistry.net/templates/shwArticle.asp?TID=3&SNID=34, Date accessed: 08/05/2006

Cerrina F, Marrian M (1996) A Path to Nanolithography. http://www.nsec.harvard.edu/AP298/Chen/path_to_nanolith.pdf, Date accessed: 03/08/2004

Chalamala BR (eds) (2000) Flat panel displays and sensors: principles, materials and processes. Symposium held April 4 - 9, 1999. Materials Research Society symposium proceedings 558. San Francisco

Childre S (2003) Recently released CMAI's 2003. World Styrene Analysis indicates an industry up-cycle is on the horizon, Chemical Market Associates (CMAI). http://www.cmaiglobal.com/news/files/2003WSA.pdf, Date accessed: 06/05/2004

CJ-Light GmbH (2003) Technische Grundlagen. CJ-Light GmbH. http://www.cj-light.de/cjlight.htm, Date accessed: 12/05/2003

Colvin VL (2002) Responsible Nanotechnology:Looking Beyond the Good News. http://www.eurekalert.org/context.php?context=nano&show=essays&essaydate=1102, Date accessed: 08/20/2006

Colvin VL (2003a) Submission to NTP Nominations Faculty /National Toxicology Programm (NIEHS). 05/19/2003: Center for Biological and Enviromental Nanotechnology

Colvin VL (2003b) Testimony before the US House of Representatives, Committee on science in regard to "Nanotechnology Research and Development Act of 2003" 04/09/2003

Crystec Technology Trading GmbH (2003) Die Herstellung von Flachbildschirmen. http://www.crystec.com/crylcdd.htm, Date accessed: 09/09/03

Dadgar A (2003) Metallorganische Gasphasenepitaxie (MOCVD). Otto von Guericke – Universität, Magdeburg. http://www.uni-magdeburg.de/ahe/lab/mocvd.html, Date accessed: 11/28/2003

David PD (1985) Clio and the economics of QWERTY. In: American Economic Review. Papers and Proceedings 75/1985:332-337

David PD (2000) Path dependence, its critics and the quest for historical economics. Working Paper, All Souls College, Oxford & Stanford University 2000. http://www-econ.stanford.edu/faculty/workp/swp00011.pdf, Date accessed: 09/18/2004

Davies JC (2005) Managing the Effects of Nanotechnology. Woodrow Wilson International Center for Scholars, Washington

Denis JH, Castor WM (1994) Styrene. In: Ullmann's Encyclopedia of Industrial Chemistry. A 25. Wiley-VCH, Weinheim

Denison RA (2005) Environmental and Safety Impacts of Nanotechnology: What Research is needed? Statement before the US House of Representatives Committee on Science, 11/17/2005, Environmental Defense, Washington http://www.environmentaldefense.org/documents/ 5136_DenisonHousetestimonyOnNano.pdf, Date accessed: 04/20/2006

Deschamps J (2000) Plasma Disply favored for large size TV screens. Paper B151: Microtec 2000, 25. - 27. September 2000. http://www.thomson.net/gb/02/pladis/pdf/Deschamps_2000_1.pdf, Date accessed: 09/28/2003

Diabaté S, Völkel K, Wottrich R (2002) Krank durch Nanopartikel. Nachrichten Biomedizin - Forschungszentrum Karlsruhe. 34 (1/2002):75-83

Die Bundesregierung (2002) Perspektiven für Deutschland: Unsere Strategie für eine nachhaltige Entwicklung. Berlin

Dierkes M, Hoffmann U, Marz L (1992) Leitbild und Technik: Zur Entstehung und Steuerung technischer Innovationen. edition sigma, Berlin

Dierkes M (ed) (1997) Technikgenese: Befunde aus einem Forschungsprogramm. edition sigma, Berlin

DIN EN ISO 14040 (1997): Umweltmanagement - Ökobilanz - Prinzipien und allgemeine Anforderungen. Beuth, Berlin

DIN - Deutsches Institut für Normung (ed) (2003) Environmental Management - Integrating Environmental Aspects into Product Design and Development. DIN-Fachbericht ISO/TR 14062. Beuth, Berlin

Dionysiou DD (2004) Environmental Applications and Implications of Nanotechnology and Nanomaterials. J. Envir. Engrg., Volume 130, Issue 7:723-724

Dockery DW, Pope CA, Xu X et al. (1993) An Association between air pollution and mortality in six US cities. N. Engl. J. Med. 329:1753-1759

Dosi G (1982) Technological paradigms and technological trajectories: a suggested interpretation of determinants and directions of technical change. Research Policy. 11 (3):147-162

Dosi G (1988) The Nature of the Innovative Process. In: Dosi G, Freeman C, Nelson R et al. (eds) Technical Change and Economic Theory. Pinter Publishers, London New York

Drexler KE (1986) Engines of Creation: The Coming Era of Nanotechnology. Foreword by Marvin Minsky. Anchor Pr., Doubleday, New York

Drexler KE, Petersen C, Pergamit G (1991) Unbounding the Future: The Nanotechnology Revolution. Morrow, New York

Drexler KE (1992) Nanosystems: Molecular Machinery, Manufacturing, and Computing. Wiley, New York

Dunphy Guzman KA, Taylor MR, Banfield JF (2006) Environmental Risks of Nanotechnology: National Nanotechnology Initiative Funding 2000-2004. In: Environ. Sci.Technol. 2006, 40 :1401-1407

Eggert P, Haid A, Wettig E (2000) Auswirkungen der weltweiten Konzentration in der Bergbauproduktion auf die Rohstoffversorgung der deutschen Wirtschaft. Duncker & Humblot, Berlin

Eickenbusch H, Hoffknecht A, Holtmannspötter D, Wagner V, Zweck A (2003) Ansätze zur technischen Nutzung der Selbstorganisation: Monitoring-Bericht. Reihe Zukünftige Technologien Nr. 48, Düsseldorf

Eigen M, Winkler R (1978) Das Spiel. Piper, München

Eikmann T, Seitz H (2002) Klein, aber oho! Von der zunehmenden Bedeutung der Feinstäube. Umweltmedizin in Forschung und Praxis. 7 (2):63-65

Environmental Defense (2005): Partnership agreement between DuPont and Environmental Defense. Washington
http://www.environmentaldefense.org/documents/5130_DuPontNanoPartnership010905.pdf, Date accessed: 04/20/2006

Engquist I (1996) Self-Assembled Monolayers. Linköping University. http://www.ifm.liu.se/Applphys/ftir/sams.html, Date accessed: 03/15/2004

Enquete-Kommission des Deutschen Bundestages Schutz des Menschen und der Umwelt (ed) (1997) Konzept Nachhaltigkeit. Zwischenbericht. Bonn

Enquete-Kommission des Deutschen Bundestages Schutz des Menschen und der Umwelt (ed) (1998) Konzept Nachhaltigkeit - Vom Leitbild zur Umsetzung (Abschlussbericht). Bonn

etc group (2002) Communiqué: No Small Matter! Nanotech Particles Penetrate Living Cells and Accumulate in Animal Organs. http://www.etcgroup.org/article.asp?newsid=356, Date accessed: 03/19/2004

Euroforum (2003) Display 2003: 8. Jahrestagung für Hersteller, Zulieferer und Anwender. http://www.euroforum.de/DATA/pdf/P15773.pdf, Date accessed: 09/11/2003

Europäische Kommission (Hrsg.) (2001) Weißbuch - Strategie für eine zukünftige Chemikalienpolitik, KOM (2001) 88 endgültig. Brüssel

Europäische Kommission (2003) Proposal for a new EU regulatory framework for chemicals. REACh (Registration, Evaluation and Authorisation of Chemicals). http://europa.eu.int/comm/enterprise/chemicals/chempol/whitepaper/REACh.htm, Date accessed: 02/11/2004

Europäische Kommission (2004) Opinion concerning Titanium Dioxide, Colipa n° S75 adopted by the SCCNFP during the 14th plenary meeting of 24 October 2000. http://europa.eu.int/comm/health/ph_risk/committees/sccp/docshtml/sccp_out 135_en.htm, Date accessed: 03/31/2004

Evident Technologies (2003) Nano Materials. http://www.evidenttech.com/, Date accessed: 01/12/2004

Farman J (2001) Halocarbons, the ozon layer and the precautionary principle. In: European Environment Agancy (ed) Late lessons from early warnings: the precautionary principle 1896-2000. EEA. 22, Copenhagen

FFU, DIW, IÖW et al. (eds) (2004) Vom Pilotmarkt zum Lead-Markt. Politikmuster der Entwicklung internationaler Märkte für Innovationen nachhaltigen Wirtschaftens. Berlin

FGL (2003) Lichtforum 40: LED - die neue Lichtquelle. Lichtforum 31: Licht und Lampen. Fördergemeinschaft Gutes Licht, Frankfurt/Main

Food and Drug Administration (1999) Sunscreen Drug Products For Over-The-Counter Human Use. Final Monograph. 21 CFR Parts 310, 352, 700, and 740 [Docket No. 78N-0038] RIN 0910-AA01. Federal Register / Vol. 64, No. 98 / Friday, May 21, 1999 / Rules and Regulations

Foresight Institute (2001) Foresight Guidelines on Molecular Nanotechnology. www.foresight.org/guidelines/current.html, Date accessed: 03/19/2004

Forrest D (1999) Regulating Nanotechnology Development (rev. 1.1). www.foresight.com/NanRev/Forrest1989.html, Date accessed: 07/03/2003

Fraunhofer IPA/IST, BAM (2004a) CVD. Beschichtungsverfahren - Klassen. http://www.schichttechnik.net/html/verfinfoklassen.cfm?SearchBy=Einzeln&Phrase=9, Date accessed: 03/05/2004

Fraunhofer IPA/IST, BAM (2004b) PVD. Beschichtungsverfahren - Klassen. http://www.schichttechnik.net/html/verfinfoklassen.cfm?SearchBy=Einzeln& Phrase=6, Date accessed: 03/05/2004

Freeman C (1995) The "National System of Innovation" in historical perspective. Cambridge Journal of Economics. 19:5-24

Funk S (2003) Entwicklung eines Anodisierverfahrens mittels Spritz/Sprühtechnik als ökologisch/ökonomisch effiziente Alternative zur Chromatierung von Aluminiumoberflächen. http://www.bwplus.fzk.de/berichte/ZBer/2003/ZBerbwd21002.pdf, Date accessed: 03/19/2004

Gabi 4 Datenbank (1999a) Ökobilanz-Datensatz für die Ethylbenzolherstellung (Niederlande). PE Europe, Leinfelden-Echterdingen

Gabi 4 Datenbank (1999b) Ökobilanz-Datensatz für die Styrolherstellung (Deutschland). PE Europe, Leinfelden-Echterdingen

Gabi 4 Datenbank (2001) Ökobilanz-Datensatz zur Herstellung von LED - Chips (Durchschnitt). PE Europe, Leinfelden-Echterdingen

Gee D, Greenberg M (2001) Asbestos: from magic to malevolent mineral. In: European Environmental Agancy (ed) (2001) Late lessons from early warnings: the precautionary principle 1896-2000. Copenhagen: European Environmental Agancy (EEA). 22

Geißler S, Harant C, Hrauda G et al. (1993) Ecodesign: Ökologische Produktgestaltung. Fibel für Anwender. Bundesministerium für Umwelt, Jugend und Familie, Wien

GEMIS 4.1 (2003) Gesamt-Emissionsmodell integrierter Systeme: Ein Programm zur Analyse der Umweltaspekte von Energie- Stoff- und Transportprozessen. Darmstadt

Gersing S (o.J.) Chemische und physikalische Oberflächenbehandlung von Aluminium. http://www.sgersing.de/pdf/SFT041.pdf, Date accessed: 03/19/2004

Gleich Av (1989) Der wissenschaftliche Umgang mit der Natur - Über die Vielfalt harter und sanfter Naturwissenschaften. Campus Forschung Bd. 601, Frankfurt/Main New York

Gleich Av (1998/1999) Ökologische Kriterien der Technik- und Stoffbewertung: Integration des Vorsorgeprinzips. Umweltwissenschaften und Schadstoff-Forschung Jg. 10, Heft 6, 1998, Jg. 11 Hefte 1 und 2 1999

Gleich Av, Rubik F (1996) Umwelteinflüsse Neuer Werkstoffe. VDI Verlag, Düsseldorf

Gleich Av (2001) Was können und was sollen wir von der Natur lernen? In: Gleich Av (ed) Bionik - Ökologische Technik nach dem Vorbild der Natur? Teubner, Stuttgart, erweiterte 2. Auflage

Gleich Av (2003) Potential ecological and health effects of nanaotechnology. Approaches to prospective technology assessment and design. In: Steinfeldt M (ed) Nanotechnology and Sustainability. Prospective Assessment of a future key technology. IÖW-SR 167-03, Berlin

Gleich Av (2004) Leitbildorientierte Technikgestaltung - Nanotechnologie zwischen Vision und Wirklichkeit. In: Böschen S, Schneider M, Lerf A (eds)

Handeln trotz Nichtwissen. Vom Umgang mit Chaos und Risiko in Politik. Industrie und Wissenschaft. Frankfurt/Main

Grahn H (2003) Das blaue Wunder - Blaue Laser aus Halbleitern. Paul - Drude - Institut für Festkörperelektronik, TU Berlin. http://www.pdi-berlin.de/semisp/dasblauewunder2.pdf, Date accessed: 10/02/2003

Grezcmiel M (2001) Entladungslampen und Umwelt. In: Bayerisches Landesamt für Umweltschutz (ed) Dokumentation der Fachtagung: Umweltrelevante Inhaltsstoffe elektrischer und elektronischer Altgeräte (EAG) bzw. Bauteile und Hinweise zu deren fachgerechten Entsorgung. Augsburg

Guston DH, Sarewitz D (2001) Real-Time Technology Assessment. Technology in Society. 23 (4):98-118

Guzman KAD, Taylor MR et al. (2006) Environmental risks of nanotechnology: National Nanotechnology Initiative funding, 2000-2004. Environmental science & technology. 40(5):1401–7.

Haase M, Kömpe K (2003) Über Nanokristalle (Halbleiter etc.). Universität Hamburg, Fachbereich Chemie. http://www.chemie.uni-hamburg.de/pc/schuelerwoche/, Date accessed: 01/12/2004

Hagen J (1999) Industrial catalysis: a practical approach. Wiley-VCH, Weinheim

Haller H (2003) LED - Info (Grundlagen, Einsatz, Zulieferer). http://www.led-info.de, Date accessed: 11/03/2003

Harsch M, Schuckert M (1996) Ganzheitliche Bilanzierung der Pulverlackiertechnik im Vergleich zu anderen Lackiertechnologien. Sachbilanzebene. Institut für Kunststoffprüfung, Stuttgart

Haum R, Petschow U, Steinfeldt M, Gleich Av (2004) Nanotechnology and Regulation within the Framework of the Precautionary Principle. IÖW-SR 173/04, Berlin

Heijungs R et al. (1992) Environmental life cycle assessment of products, backgrounds & guide. Centre of Environmental Science, Leiden

Hellige HD (1996) Technikleitbilder als Analyse-, Bewertungs- und Steuerungsinstrumente. Eine Bestandsaufnahme aus informatik- und computerhistorischer Sicht. In: Hellige HD (ed) Technikleitbilder auf dem Prüfstand. Leitbild-Assessment aus Sicht der Informatik- und Computergeschichte. Sigma, Berlin

Hemmelskamp J (ed) (2001): Forschungsinitiative zu Nachhaltigkeit und Innovation. ökom, München

Herrmann WA (2000) Zukunftstechnologie Katalyse. In: Felcht UH (ed) Chemie. Eine reife Industrie oder weiterhin Innovationsmotor? Blazek und Bergmann, Frankfurt/Main

Hett A (2004) Nanotechnologie. Kleine Teile-Große Zukunft. Swiss Reinsurance Company, Zürich

Hoet M et al. (2004) Health impact of Nanomaterials. Letter to the Editor. Nature Biotechnology. 22 (1):19

Höhr D, Steinfartz Y, Schins RPF, Knaapen AM, Martra G, Fubini B, Borm PJA (2002) The surface area rather than the surface coating determines the acute inflammatory response after instillation of fine and ultrafine TiO2 in the rat. Int. J. Hyg. Environ. Health.205:239-244.

Holling CS (1994) New Science and New Investments for a Sustainable Biosphere. In: Jansson A, Hammer M, Costanza R (eds) Investing in Natural Capital - The Ecological Economics Approach to Sustainability. Island Press, Washington

Howard VC (2004a) Presentation to European Parliament. Brussels

Howard VC (2004b) Nano-particles and toxicity. Unpublished Paper

Huber J (2001) Ökologische Konsistenz. Zur Erläuterung und kommunikativen Verbreitung eines umweltinnovativen Ansatzes. In: Umweltbundesamt (ed) Perspektiven für die Verankerung des Nachhaltigkeitsleitbildes in der Umweltkommunikation, UBA-Berichte 4/01, Erich Schmidt, Berlin, pp 80-100

Huber J (2004) New Technologies and Environmental Innovation. Edward Elgar, Cheltenham

Hübner H (2002) Integratives Innovationsmanagement - Nachhaltigkeit als Herausforderung für ganzheitliche Erneuerungsprozesse. Erich Schmidt, Berlin

Hunt CE (ed) (2001) Proceedings of the 14th International Vacuum Microelectronics Conference. Davis

ICON (2006) 01.04.2006. http://icon.rice.edu, Date accessed: 04/15/2006

IGVT (2003) Nanopartikel für die Molekulare Erkennung. Institut für Grenzflächenverfahrenstechnik, Universität Stuttgart. http://www.uni-stuttgart.de/igvt/forschung/molekularErkenn-deu.htm, Date accessed: 03/08/2004

IMST (2002) Dünne Funktionsschichten aus dem Sol-Gel Verfahren. Institute for Materials & Surface Technology, FH Kiel. http://www.wpr.fh-kiel.de/research/solgel/solgel.htm, Date accessed: 03/08/2004

Information Society Technologies (2003) Carbon nanotubes for large area displays (Canadis Project, IST-1999-20590). http://canadis.crm-paris.com/Pagecanadis.html, Date accessed: 09/20/2003

International Risk Governance Council (2005) Survey on Nanotechnology Governance. Volume D: The Role of NGO's. IRGC Working Group on Nanotechnology. Chair: Roco MC, Member IRGC; P.M.: Litten E. Geneva

International Risk Governance Council (IRGC) (Renn O, Roco MC) (2006) IRGC White Paper No. 2: Nanotechnology Risk Governance. http://www.irgc.org/irgc/_b/contentFiles/IRGC_white_paper_2_PDF_final_version.pdf, Date accessed: 12/25/2006

IOM - Institute for Occupational Medicin (ed) (2005) A scoping study to identify hazard data needs for addressing the risks presented by nanoparticles and nanotubes. Prepared by Tran CL, Donaldson K, Stones V, Fernandez T, Ford A, Christofi N, Ayres JG, Steiner M, Hurley JF, Aitken RJ, Seaton A, Edinburgh

IOM - Institute for Occupational Medicin (ed) (2006) A scoping study to identify gaps in environmental regulation for the products and applications of nanotechnologies. Prepared by Chaudhry Q, Blackburn J, Floyd P, George C, Nwaogu T, Boxall A, Aitken R, Edinburgh

IPMS (2003) OLED. Fraunhofer Institut Photonische Mikrosysteme, Dresden. http://www.ipms.fhg.de/products/index_09.html, Date accessed: 30.03.2004

ISIF (2001) Information on the Styrene Industry Worldwide. International Styrene Industry Forum (ISIF). http://www.styreneforum.org/industry_index.html, Date accessed: 03/22/2004

Jefferson DA, Tilley EEM (1999) The structural and physical chemistry of nanoparticles. In: Maynard RL, Howard CV (eds) Particulate matter. Properties and effects upon health. Cromwell Press, Trowbridge

Jelinek TW (1997) Oberflächenbehandlung von Aluminium. Leuze, Saulgau/Württ

Jelinski L (1999) Biologically Related Aspects of Nanoparticles, Nanostructured Materials and Nanodevices. National Science and Technology Council. http://www.wtec.org/loyola/nano/07_02.htm, Date accessed: 03/08/2004

Jensen KL (ed) (2001) Electron emissive materials, vacuum microelectronics and flat panel displays. Symposium held April 25 - 27, 2000. Materials Research Society symposium proceedings 621. San Francisco

Jonas H (1985) Warum die moderne Technik ein Gegenstand für die Ethik ist. In: Jonas H (ed): Technik, Medizin, Ethik. Zur Praxis des Prinzips Verantwortung. Insel, Frankfurt/Main

Joy B (2000) Why the future doesn't need us. Wired Magazine. http://www.wired.com/wired/archive/8.04/joy_pr.html, Date accessed: 03/19/2004

Jüstel T, Feldmann C, Ronda CR (2000) Leuchtstoffe für aktive Displays. Physikalische Blätter. 56 (9):55-58

Kemp R (1997) Environmental Policy and Technical Change. A Comparison of the Technological Impact of Policy Instruments, Cheltenham, Edward Elgar

Ketteler G (2002) Präparation und Charakterisierung von epitaktischen Oxidfilmen für modellkatalytische Untersuchungen. Dissertation, FU Berlin. http://www.diss.fu-berlin.de/2003/12/, Date accessed: 03/22/2004

KEVAG (2003) Die Energiesparlampe. Koblenzer Elektrizitätswerk und Verkehrs - Aktiengesellschaft. http://www.kevag.de/strom/get_ag/energiesparlampe.php, Date accessed: 11/18/2003

Klöpffer W, Curran MA, Frankl P, Heijungs R, Köhler A, Olsen SI (2007) Nanotechnology and Life Cycle Assessment. Synthesis of Results Obtained at a Workshop, Washington, DC, 2–3 October 2006. http://www.nanotechproject.org/111/32007-life-cycle-assessment-essential-to-nanotech-commercial-development, Date accessed: 03/21/2007

Köhler M (2001) Nanotechnologie. Eine Einführung in die Nanostrukturtechnik. Wiley-VCH, Weinheim

Kowalsky W, Benstem T, Böhler A et al. (1999) Organic Electroluminescent Devices Advances. Solid State Physic. 39:91-100

Kowol U (1998) Innovationsnetzwerke. Technikentwicklung zwischen Nutzungsvisionen und Verwendungspraxis. Wiesbaden

KPMG (1999) Unternehmensleitbilder in deutschen Unternehmen, eine Untersuchung von KPMG in Zusammenarbeit mit dem Lehrstuhl für Unternehmensführung an der Universität Erlangen-Nürnberg. Frankfurt Nürnberg.

Krämer M (2002) Feinste Partikel aus dem Laserverdampfungsreaktor. Elements, Degussa ScienceNewsletter. (01/2002):9-11
Kreyling WG, Semmler M, Möller W (2004) Dosimetry and toxicology of ultrafine particles. Aerosol Med, 17:140-52
Krost A, Dadgar A (2002) Optoelektronik auf Silizium - eine Herausforderung für die Halbleiterphysik. Magdeburger Wissenschaftsjournal, 1/2002
Kuhn T (1975) Neue Überlegungen zum Begriff des Paradigma. In: Kuhn T (ed) Die Entstehung des Neuen. Studien zur Struktur der Wissenschaftsgeschichte. Suhrkamp Wissenschaft, Frankfurt/Main
Kühner G (1999) What is Carbon Black®? Firmenschrift Degussa AG, Frankfurt/Main
Kunzli N, Kaiser R, Medina S (2000) Public-health impact of outdoor and traffic-related air pollution: a European assessment. Lancet. 356:795-801
Laval JM, Chopineau J, Thomas D (1995) Nanotechnology: R&D challenges and opportunities for application in biotechnology. Tibtech. 13:474-481
Lecoanet H, Wiesner MR (2004) Assessment of the mobility of nanomaterials in groundwater acquifers. Conference: Nanotechnology and the Environment. Anaheim, March28-April1 2004, http://oasys2.confex.com/acs/227nm/techprogram/S13728.HTM, Date accessed: 09/18/2004
Lekas D (2005a) Analysis of Nanotechnology from an Industrial Ecology Perspective Part I: Inventory & Evaluation of Life Cycle Assessments of Nanotechnologies. http://www.nanotechproject.org/15/analysis-of-nanotechnology-from-an-industrial-ecology-perspective, Date accessed: 03/18/2006
Lekas D (2005b) Analysis of Nanotechnology from an Industrial Ecology Perspective. Part II: Substance Flow Analysis Study of Carbon Nanotubes. http://www.nanotechproject.org/15/analysis-of-nanotechnology-from-an-industrial-ecology-perspective, Date accessed: 03/18/2006
Lieb M, Hildebrand B (1982) Styrol. In: Ullmanns Encyklopädie der technischen Chemie. Wiley-VCH, Weinheim
Lohmann A (1997) Bildschirme und Displays. In: Fachbericht Neue Rechnertechnologien: Fachhochschule Augsburg, Fachbereich Informatik, pp 66-85
Lloyd S, Lave L (2003) Life Cycle Economic and Environmental Implications of Using Nanocomposites in Automobiles. Environmental Science & Technology. 37(15):3458-66.
Lloyd S, Lave L, Matthews S (2005) Life Cycle Benefits of Using Nanotechnology to Stabilize Platinum-Group Metal Particles in Automotive Catalysts. Environmental Science & Technology. Published online January 15.
Lueder E (2001) Liquid Crystal Displays: Addressing Schemes and Electro-Optical Effects. John Wiley & Sons, New York
Luther W (2004) Industrial applikation of nanomaterial - chances and risks. Reihe Zukünftige Technologien, Bd. 54, Düsseldorf
Macroubie J (2005) Informed Public Perceptions of Nanotechnology and Trust in Government. Woodrow Wilson International Center for Scholars, Washington DC

Maichin R (2002) Statuserhebung der Nanotechnologieforschung an akademischen Einrichtungen in der Steiermark. Diplomarbeit Technische Universität, Graz http://www.tvtut.tugraz.at/tvtut/nanotechnologie/Nanotechnologieforschung.pdf, Date accessed: 03/05/2004

Mambrey P, Paetau M, Tepper A (1995) Technikentwicklung durch Leitbilder. Neue Steuerungs- und Bewertungsinstrumente. Campus, Frankfurt/Main

Mani J (1994) Eine Ökobilanz von Glühlampe und Energiesparlampe. Büro '84, Bern

Marz L, Dierkes M (1994) Leitbildprägung und Leitbildgestaltung. Zum Beitrag der Technikgenese-Forschung für eine prospektive Technikfolgen-Regulierung. In: Bechmann G, Petermann T (eds) Interdisziplinäre Technikforschung. Genese, Folgen, Diskurs. Campus, Frankfurt/Main, pp 35-71

Matje A (1996) Unternehmensleitbilder als Führungsinstrument. Komponenten einer erfolgreichen Unternehmensidentität. Gabler, Wiesbaden

Maturana HR, Varela FJ (1987) Der Baum der Erkenntnis. Bern

Maximova N (2002) Partialdehydrierung von Ethylbenzol zu Styrol an Kohlenstoffmaterialien. Dissertation, TU Berlin

Maynard AD (2006) Nanotechnology: assessing the risks. Nanotoday 1(2): 22-33

McDonough W, Braungart M (2002) Cradle to Cradle - Remaking the Way We Make Things. North point Press, New York

Merck KGaA (2000) Stellungnahme des Umweltbundesamtes über die Ökotoxikologie von Flüssigkristallen in Flüssigkristallanzeigen. http://www.merck.de/servlet/PB/menu/1104220/index.html, Date accessed: 03/31/2004

Merkle RC (1994) Self-replicating systems and low cost manufacturing. In: Welland ME, Gimzewski JK (eds) The Ultimate Limits of Fabrication and Measurement. Kluwer, Dordrecht, pp 25-32

Mestl G (2004) Persönliche Mitteilung. Nanoscape AG, München

Meyer-Krahmer F (1997) Umweltverträgliches Wirtschaften. Neue industrielle Leitbilder, Grenzen und Konflikte. In: Blättel-Mink B, Renn O (eds) Zwischen Akteur und System. Die Organisation von Innovation. Westdeutscher Verlag, Opladen

Mink E, Rzepka M (1995) Maßnahmen zur Emissionsminderung bei nicht genehmigungsbedürftigen Anlagen. In: UTECH (ed) Anwendung umweltfreundlicher Lacksysteme. Berlin

Mounier E (2002) Micro & Nanotechnologies for displays and data storage in portable systems: a quick overview: Yole Developpement. http://colossalstorage.net/eloy_3c.ppt, Date accessed: 09/28/2003

MPG (2002) Nanozwiebeln bringen Styrol-Synthese auf Trab. Max-Planck-Gesellschaft http://www.mpg.de/bilderBerichteDokumente/dokumentation/pressemitteilungen/2002/pri0299.htm, Date accessed: 09/28/2003

National Science and Technology Council (1999) Nanotechnology. Shaping the World Atom by Atom. Washington DC

NEDO: New Energy and Industrial Technology Development Organization. http://www.nedo.go.jp/nanoshitsu/project/loadmap_eng.pdf.

Nelson RR (ed) (1993) National Systems of Innovation. A comparative analysis. University Press, Oxford

New Energy and Industrial Technology Development Organization (2000) Road Map of Technology. http://www.nedo.go.jp/nanoshitsu/project/loadmap_eng.pdf, Date accessed: 09/23/2004

New Scientist (2003) Carbon Nanotubes show drug delivery promise. http://www.newscientist.com/news/news.jsp?id=ns99994485, Date accessed: 03/31/2004

Niesen TP, Aldinger F (2001) Biomineralisation & Biomimetische Materialherstellung. In: Gleich Av (ed) Bionik - Ökologische Technik nach dem Vorbild der Natur? Teubner, Stuttgart

NNI (2000) National Nanotechnology Initiative. The Initiative and its Implementation Plan. Washington DC

Nocula C, Olbrich S (2003) Zur Diskrepanz der Auflösung verschiedener Präsentationsmedien: Aktuelle Entwicklungen der Bildschirmtechnologien: Universität Hannover. http://www.rtb-nord.unihannover.de/onlinedokumente/abschlussbericht/p5_1a.pdf, Date accessed: 06/29/2004

Nordmann A (2003) 'Shaping the World Atom by Atom': Eine nanowissenschaftliche WeltBildanalyse. In: Grunwald A (ed.) Technikgestaltung zwischen Wunsch und Wirklichkeit. Springer, Berlin Heidelberg New York

Nordmann A (2003) Molecular Disjunctions. DP Discovering the Nanoscale.

Nordmann A (2004) New Space for Old Cosmologies: Imaging the World for Nanoscience. Conference on Imaging and Imagining NanoScience and Engineering, Columbia SC, March 3-7

Oberdörster G, Sharp Z, Atudorei V et al. (2002) Extrapulmonary translocation of ultrafine carbon particles following whole-body inhalation exposure of rats. Toxicology and Environmental Health A. 65 (20):1531-43

Oberdörster G, Sharp Z, Atudorei V, Elder A, Gelein R, Kreyling W, Cox C (2004) Translocation of inhaled ultrafine particles to the brain. Inhal Toxicol. Jun;16(6-7):437-445

Oberdörster G, Oberdörster E, Oberdörster J (2005) Invited review: Nanotechnology: an emerging discipline evolving from studies of ultrafine particles. Environ Health Perspect. 2005 Jul;113(7):823-839

Oberdörster G, Maynard A, Donaldson K, Castranova V, Fitzpatrick J, Ausman K, Carter J, Karn B, Kreyling W, Lai D, Olin S, Monteiro-Riviere N, Warheit D, Yang H (2005) Principles for characterizing the potential human health effects from exposure to nanomaterials: elements of a screening strategy. A report from the ILSI Research Foundation/Risk Science Institute Nanomaterial Toxicity Screening Working Group. Particle and Fibre Toxicology 2005, 2:8

OECD (2006) Report of the OECD Workshop on the Safety of Manufactured Nanomaterials. Building Co-operation, Co-ordination and Communication .Washington D.C., United States, 7th-9th December 2005. Series on the Safety of Manufactured Nanomaterials No. 1; Environment Directorate, Joint Meeting of the Chemicals Committee And The Working Party on Chemicals, Pesticides And Biotechnology. ENV/JM/MONO(2006)19

Omercon (o.J.) Organische Metalle – Leitfähige Polymere. http://www2.ormecon.de/Environment/Factor10.html, Date accessed: 06/29/2006

Osterwalder N, Capello C, Hungerbühler K, Stark WJ (2006) Energy consumption during nanoparticle production: How economic is dry synthesis? Journal of nanoparticle Research. 8:1-9

Paschen H, Coenen C, Fleischer T, Grünwald R, Oertel D, Revermann C (2003) TA-Projekt Nanotechnologie. Arbeitsbericht Nr. 92., Büro für Technikfolgen-Abschätzung beim Deutschen Bundestag (TAB), Berlin

Perrelin C (2006) Nanotechnology Developments Prompt Policy Questions. http://usinfo.state.gov/gi/Archive/2006/Feb/16-601911.html, Date accessed: 06/18/2006

Pescovitz D (2004) Nano, Bio converge to provide nanotech link. Small Times 08/06/2004, http://www.smalltimes.com/document_display.cfm?section_id=76&document_id=8188, Date accessed: 09/18/2004

Petschow U, Rosenau J, Weizsäcker EUv (2005) Governance and Sustainability. New Challenges for States, Companies and Civil Society

Pflücker F, Wendel V, Hohenberg H et al. (2001) The Human Stratum corneum Layer: An effective Barrier against dermal Uptake of Different Forms of topically applied micronised Titanium Dioxide. Skin Pharmacology and Applied Skin Physiology. 14 (1):92-97

Pope CA, Dockery DW (1999) Epidemiology of particle effects. In: Holgate ST, Samet JM, Koren HS, Maynard RL (eds) Air Pollution and Health. Academic Press, San Diego, pp 673-705

Popular Mechanics (2002) Flourescent Tube. http://www.popularmechanics.com/popmech/homei/9110HIWAM.html, Date accessed: 12/10/2003

Pressetext (2004) LG Philips mit Rekord-OLE-Display. 20.10.2004, http://www.pressetext.de/pte.mc?pte=041020021, Date accessed: 06/05/2006

Quella F (ed) (1998) Umweltverträgliche Produktgestaltung. Siemens Fachpublikation. Wiley-VCH, Weinheim

Rahman Q, Lohani M, Dopp E et al. (2002) Evidence That Ultrafine Titanium Dioxide Induces Micronuclei and Apoptosis in Syrian Hamster Embryo Fibroblasts. Environmental Health Perspectives. 110 (8).

Rejeski D (2005) Environmental and Safety Impacts of Nanotechnology: What Research is Needed? Testimony. United States House of Representatives, Committee on Science, 11/17/2005. www.nanotechproject.com/index.php?s=file_download&id=15, Date accessed: 06/06/2006

Rejeski D (2003) Welcome to the next industrial revolution. Presentation given to the National Science Foundation. http://es.epa.gov/ncer/publications/nano/pdf/RejeskiNSF(9.15.PDF, Date accessed: 06/20/2006

Reynolds A (2002) Forward to the future: Nanotechnology and Regulatory Policy.

Reynolds GH (2001) Environmental Regulation of Nanotechnology: Some Preliminary Observations: Environmental Law Institute.
www.foresight.org/impact/31.10681.pdf, Date accessed: 03/19/2004

Rickmeyer C (2002) Penetrationseigenschaften von beschichtetem mikrofeinem Titandioxid. Medizinische Fakultät Charité, Humboldt Universität, Berlin

Riebeek H (2003) Bright Quantum Light - Huge jumps in efficiency could make quantum dot LEDs the future of flat - panel displays. IEEE - Institute of Electrical and Electronic Engineers, Web - Only - News.
http://www.spectrum.ieee.org/WEBONLY/wonews/feb03/quantum.html, Date accessed: 01/11/2004

Rip A, Misa TJ, Schot JW (eds) (1995) Managing Technology in Society. The Approach of Constructive Technology Assessment. London.

Robinson C (2003) "Visualizing Nanotechnology", Conference on Discovering the Nanoscale II, Darmstadt, Germany, October 9-12, 2003

Robinson C (2004) "What does it mean to see?". Conference on Imaging and Imagining NanoScience and Engineering, Columbia SC, March 3-7, 2004

Roco MC (2002) The Future of National Nanotechnology Initiative. Presentation given to National Science and Technology Council.
http://www.nsf.gov/home/crssprgm/nano/roco_aiche_48slides.pdf, Date accessed: 03/19/2004

Roco MC, Bainbridge W (eds) (2002) Converging Technologies for Improving Human Performance: Nanotechnology, Biotechnology, Information Technology, and Cognitive Science. The National Science Foundation, Arlington, Virginia

Rohden D (2001) CMAI publishes results of 2001 World Styrene Analysis and 2001 World Polystyrene/EPS Analysis: Chemical Market Associates Inc. (CMAI). http://www.cmaiglobal.com/news/files/WSA-WPS_EPS2001.pdf, Date accessed: 03/22/2004

Rossi M (2000) Atomlithographie. Hauptseminar SS2000 Ultrakalte Atome und Atomoptik: Universität Stuttgart. http://www.physik.uni-stuttgart.de/institute/pi/5/lehre/hauptseminar2000/atomlithographie/atomlithographie_druck.html, Date accessed: 03/12/2004

Rössler A, Skillas G, Pratsinis SE (2001) Nanopartikel - Materialien der Zukunft. Chemie in unserer Zeit. 35 (1):32-41

Royal Academy of Engineering (2003) Nanotechnology: views of Scientists and Engineers. http://www.nanotec.org.uk/SEworkshopReport1.pdf, Date accessed: 03/29/2004

Rubahn HG (2002) Nanophysik und Nanotechnologie. Teubner, Stuttgart

Samsung (2003) Persönliche Auskunft von Chun Gyoo Lee. Principal Researcher Corporate R&D Center Samsung SDI.

Samsung (2005) SAMSUNG Electronics Develops World's First 40-inch a-Si-based OLED for Ultra-slim, Ultra-sharp Large TVs. Pressemitteilung des Unternehmens,
http://samsung.com/PressCenter/PressRelease/PressRelease.asp?seq=20050519_0000123644, Date accessed: 05/08/2006

Sauer D, Lang C (eds) (1999) Paradoxien der Innovation. Perspektiven sozialwissenschaftlicher Innovationsforschung. Campus, Frankfurt/Main New York

Schoen M (2002) Katalytische Untersuchungen zur Styrolsynthese an Eisenoxidkatalysatoren. Ruhr Universität, Bochum

Schummer J (o.J.) Projektentwurf „Philosophische Analysen der Nanoforschung" http://www.cla.sc.edu/cpecs/nirt/grants/Heisenberg_proposal.pdf, Zugrifsdatum: 09/18/2004

Schwarzer HC (2001) Erzeugung von Nanopartikeln durch Fällung. TU München. http://www.lfg.mw.tum.de/persons/hschwarzer/schwarzer.html, Date accessed: 03/08/2004

Scientific Committee On Emerging And Newly Identified Health Risks (SCENIHR) (2005) Opinion on The appropriateness of existing methodologies to assess the potential risks associated with engineered and adventitious products of nanotechnologies. Adopted by the SCENIHR during the 7th plenary meeting of 28-29 September 2005. SCENIHR/002/05. European Commission Health & Consumer Protection Directorate-General, Directorate C - Public Health and Risk Assessment C7 - Risk assessment

Scott CL, Malliaras GG, Bozano L et al. (2000) The Physics of organic light emitting devices. In: Materials Research Society (ed) Flat panel displays and sensors: principles, materials and processes. Symposium held April 4 - 9, 1999. Symposium proceedings Vol. 558. San Francisco, pp 499 - 505

Senatsverwaltung für Stadtentwicklung Berlin (2003) Elektroschrott - Daten, Fakten, Hintergründe zur Abfallwirtschaft.
http://www.stadtentwicklung.berlin.de/umwelt/abfallwirtschaft/de/elektroschrott/index.shtml, Date accessed: 11/24/2003

Siemens AG (2003) News Desk: Leuchtdioden mit Heizung auf der Rollbahn (26.11.2003), Leuchtdioden entlasten Umwelt (05.11.2003), Leuchtdioden als Autoscheinwerfer (05.09.2003), Leuchtdioden bald allgegenwärtig (05.07.2003), Kombinatorik erleichtert Suche nach neuem LED - Material (06.05.2003).
http://w4.siemens.de/de2/html/press/newsdesk_archive/index.html, Date accessed: 12/04/2003

Singh A, Markowitz M, Chow G (1995) Materials Fabrication via Polymerizable Self-Organized Membranes: An Overview. NanoStructured Materials. 5 (2), pp 141-153

Small Times (2002) 02/12/2002. http://www.smalltimes.com.

Small Times (2006) 04/14/2006.
http://www.smalltimes.com/document_display.cfm?document_id=11289, Date accessed: 05/05/2006

Smalley RE (1999) Prepared Written Statement and Supplemental Material. In: Roco MC, Williams S, Alivisatos P (eds): Vision for Nanotechnology Research and Development in the Next Decade - Nanotechnology Research Directions: IWGN Workshop Report.

SNL (2003): Sandia researchers use quantum dots as a new approach to solid - state lightning. Sandia National Laboratories. http://www.sandia.gov/newscenter/news-releases/2003/elect-semi-sensors/quantum.html, Date accessed: 01/14/2004

Socolof ML, Overly JG, Kincaid LE et al. (2001) Desktop Computer Displays: A Life-Cycle Assessment. Endbericht für United States Environmental Protection Agency, University of Tennessee Center for Clean Products an Clean Technologies.

Steinfeldt M (ed) (2003) Nanotechnology and Sustainability. Prospective assessment of a future key technology. IÖW-SR 167/03, Berlin

Steinfeldt M, Petschow U, Hirschl B (2003) Anwendungspotenziale nanotechnologiebasierter Materialien. Teilgebiet 2: Analyse ökologischer, sozialer und rechtlicher Aspekte. IÖW-SR 169/03, Berlin

Steinfeldt M (2003) Nanotechnologie: Bewertungskriterien und Gestaltungsansatz. Die Nachhaltigkeit der kleinsten Dinge. In: Ökologisches Wirtschaften 6/03:20-21

Steuber F (2000) Untersuchungen zur Effizienz, Farbabstimmung und Degradation von organischen Elektrolumineszenz-Anzeigen. Berichte aus der Halbleitertechnik. Shaker, Aachen

Tang CW, Van Slyke SA (1987) Organic electroluminescent diodes. Applied Physics Letters. 51 (12):913-915

Tang YS, Sotornayor-Torres CM, Wilkinson CDW (2003) Si - SiGe Quantum Dot - based Light Emitting Diodes. Nanoelectronic Research Center, Glasgow University. http://www.elec.gla.ac.uk/groups/nanospec/SiGe.html, Date accessed: 01/13/2004

Tannas LE (1985) Flat panel displays and CRTs. VanNostrand, New York

The Royal Society, Royal Academy of Engineers (2004): Nanoscience and nanotechnologies: opportunities and uncertainties. Final Report July 2004 http://www.nanotec.org.uk/finalReport.htm, Date accessed: 09/18/2004

Theis D (2000) Displays - Schlüsselkomponenten der Informationsgesellschaft. Physikalische Blätter. 56 (9):59-63

Thimm U (1999) Umweltfreundliche Halbleiterproduktion - Aus Forschern werden Unternehmer. Marburger Uni - Journal. (1/1999):26-29

Tinkle SS, Antonini JM, Rich BA et al. (2003) Skin as a Route of Exposure and Sensitization in Chronic Beryllium Disease. Environmental Health Perspectives. 111 (9)

Tomson M, Colvin VL, Cheng X et al. (2003) Adsorption/Desorption of Pollutants to Nanoparticles. Center for Biological and Environmental Nanotechnology Civil and Environmental Engineering and Chemistry Departments Rice University. September 15, 2003. Interagency Workshop: Nanotechnology and the Environment: Applications and Implications

Uchino T, Tokunaga H, Ando M et al. (2002) Quantitative determination of OH radical generation and its cytotoxicity induced by TiO2-UVA treatment. Toxicology in Vitro. 16:629-635

Umweltbundesamt (UBA) (1999) Technische Optionen zur Verminderung der Verkehrsbelastungen. Texte-Reihe des UBA Nr. 33/99, Berlin

Umweltbundesamt (UBA) (ed) (2000) Was ist EcoDesign? Ein Handbuch für ökologische und ökonomische Gestaltung. Berlin

Umweltbundesamt (UBA) (2006) Nanotechnik: Chancen und Risiken für Mensch und Umwelt. Berlin. http://umweltbundesamt.de/uba-info-presse/hintergrund/nanotechnik.pdf, Date accessed: 01/13/2007

UN (2004) Environmental Requirements and Market Access for Developing Countries. Paper of the UN Conference on Trade and Development, Sao Paulo http://www.mvwebsolutions.com/download/td_xi_bp_1%20presessional.doc?PHPSESSID=f4efcd641d6d7915e39881be1929a278, Date accessed: 06/19/2006

U.S. Environmental Protection Agency (2005) Nanotechnology White Paper - External Review Draft. Prepared for the U.S. Environmental Protection Agency by members of the Nanotechnology Workgroup, a group of EPA's Science Policy Council. December 2, 2005

U.S. House of Representatives Committee on Science (2005) Environmental and Safety Impacts of Nanotechnology: What Research is Needed? Hearing Thursday, November 17, 2005

Van Ooij WJ, Zhu D, Palanivel V et al. (2002) Potential of Silanes for chromate replacement in metal finishing industries. University Of Cincinnati. http://www.eng.uc.edu/~wvanooij/SILANE/pot_Chromate_repl.pdf, Date accessed: 03/19/2004

VCI (2003) Chemiewirtschaft in Zahlen. Verband der Chemischen Industrie, Frankfurt/Main

VDI (ed) (2002) VDI-Richtlinie 2243: Recyclingorientierte Produktentwicklung.

VDI-Nachrichten (2003) VDI-Nachrichten. 33, 2003-10-08

VDI-Technologiezentrum (ed) (2002) Technologiefrüherkennung Technologieanalyse Nanobiotechnologie I: Grundlagen und Anwendungen molekularer, funktionaler Biosysteme. Reihe Zukünftige Technologien Nr. 38, Düsseldorf

VDI Technologiezentrum Zukünftige Technologien Consulting (ed) (2003) Ansätze zur technischen Nutzung der Selbstorganisation. Monitoring-Bericht. Reihe Zukünftige Technologien Nr. 48, Düsseldorf

VDI Technologiezentrum Zukünftige Technologien Consulting (ed) (2004) Kontrollierte Selbstorganisation für zukünftige technische Anwendungen. Fachgespräch, Analyse, Ausblick. Reihe Zukünftige Technologien Nr. 55, Düsseldorf

Vision 2020 (2002) Nanomaterials and the Chemical Industry - R&D Roadmap Workshop. Preliminary Results. Workshop held on September 30, October 1 and 2, 2002. Chemical Industry: Vision 2020 - Technology Partnership

Vista Verde (2002) Schwermetalle: Aus dem Bergbau in die Flüsse. 27.09.2002, http://www.vistaverde.de/news/Wissenschaft/0209/27_bergbau.htm, Date accessed: 05/05/2006

Vollrath F (1992) Die Seiden und Netze von Spinnen. Spektrum der Wissenschaft, Mai 1992:82-89

Vollrath F, Knight DP (2001) Liquid crystalline spinning of spider silk. Nature. 410:541-548

Volz S, Olson W (2004) Life Cycle Assessment and Evaluation of Environmental Impact of Carbon Nanofiber Reinforced Polymers. Submitted to the Journal of Industrial Ecology on August 2, 2004.

Vossloh Schwabe Deutschland GmbH (2003) Elektromagnetische Verträglichkeit von Leuchten und Leuchtenzubehör; Topmagnetic Today - Neueste Erkenntnisse aus Technik und Normen; Energiesparende Leuchtstofflampen. http://www.vossloh-schwabe.com/ger/news/index.php, Date accessed: 12/09/2003

Wagner G et al. (o.J.) Organic-Inorganic Sol-Gel Coatings as Corrosion Protection for Metals. Nano Tech Coatings GmbH, Tholey. http://www.ebs.eurasia.de/download/, Date accessed: 09/01/2003

Wagner G (o.J.) Nanotechnologie in modernen Beschichtungsmaterialien aus lackchemischer Sicht. http://www.phaenomen-farbe.de/pf_812_forschung_nano-2.htm, Date accessed: 03/19/2004

Wardak A (2003) Nanotechnology & Regulation. A Case Study using the Toxic Substance Control Act (TSCA) Foresight and Governance Project. Woodrow Wilson School

Warheit DB, Laurence BR, Reed KL et al. (2004) Comparative Pulmonary Toxicity Assessment of Single-wall Carbon Nanotubes in Rats. Toxicological Sciences. 77:117-125

Wehling P (2001) Jenseits des Wissens? Wissenschaftliches Nichtwissen aus soziologischer Perspektive. Zeitschrift für Soziologie. 30:465-484

Wehling P (2002) Rationalität und Nichtwissen. (Um-)Brüche gesellschaftlicher Rationalisierung. In: Karafyllis N, Schmidt J (eds) Zugänge zur Rationalität der Zukunft. Metzler, Stuttgart

Wiedemann P, Karger C, Clauberg M (2002) Risikofrüherkennung im Bereich Umwelt und Gesundheit, Aktionsprogramm, Umwelt und Gesundheit, Teilvorhaben 9, Umweltforschungsplan F+E-Vorhaben 200 61 218/09. Jülich

WKO (2003) Energiewirtschaft und Energietechnik: Energiesparende Beleuchtung: Wirtschaftskammer Oberösterreich. http://www.wko.at/ooe/energie/Branchen/Licht/licht.htm, Date accessed: 11/24/2003

Wuelfert S (2000) Lichtquellen. http://surf.agri.ch/wuelfert/LECTURE/PHYSICS/COLOR/light_sources/color_light_sources.html, Date accessed: 09/26/2003

Würdinger E, Roth U, Wegener A et al. (2000) Kunststoffe aus nachwachsenden Rohstoffen: Vergleichende Ökobilanz für Loose-fill-Packmittel aus Stärke bzw. Polystyrol. Endbericht: Institut für Energie- und Umweltforschung (IFEU). http://www.ifeu.de/englisch/agri/seiten/a_ref3.htm, Date accessed: 03/22/2004

ZVEI (2003) Verschiedenste Informationen der Internetseite z.B. LED in der Allgemeinbeleuchtung: Zentralverband Elektrotechnik und Elektrotechnikindustrie e.V. http://www.zvei.org/, Date accessed: 09/26/2003

Index

Actors 183, 185, 196, 197, 200
Aqueous manufacturing method 187
Asbestos 20, 21, 22

BfR 201, 202
Biomimetic synthesis 190, 191, 194

CFCs 18, 20, 21, 26, 29, 34
Classification 9, 15, 29, 81, 172, 178, 179, 207, 210, 218, 219
Communication 12, 14, 32, 183, 187, 195, 196, 199, 203, 211, 212, 220, 221
  upstream 202
Communication technology 112
Constructive technology assessment 13, 14, 33, 196
Cooperation VIII, XV, 12, 120, 142, 183, 184, 191, 197, 203, 211
Cooperative approaches 184

Design development 182, 197
Design process 183, 195, 196, 197, 204
Dissemination VIII, 21, 34, 196, 201, 211

Ecological adaptiveness 189
Energy efficiency 8, 9, 66, 109, 111, 116, 117, 130, 135, 136, 139, 157, 159, 204
Engineering results assessment 1, 13, 17, 25, 38, 184
EPA 120, 170, 208, 209, 210

EU chemical regulatory approach 210
Evolutionary approvedness 189, 190

Flame-assisted deposition 4, 46, 48, 59, 176
Foresight guidelines 182, 198
Foresight Guidelines 188, 209
Fullerenes 31, 161, 164, 170

Gas-phase deposition 46
Geneva 199
Governance 199
Governance problem 12, 183

Hazards 2, 3, 5, 9, 13, 14, 22, 25, 38, 110, 136, 162, 163, 172, 175, 178, 179, 182, 195, 196, 198, 199, 203, 207, 213, 217

ICON 199, 201, 212
Ingenuity 190
ISO 28, 200, 204, 211, 212, 213

Leitbild IX, 1, 10, 11, 12, 13, 14, 17, 33, 36, 37, 38, 181, 184, 186, 188, 189, 190, 196, 197, 215
Life cycle assessment IX, XV, 1, 2, 5, 8, 13, 17, 27, 28, 30, 41, 65, 190, 197, 215, 216, 221
  displays 112, 120, 130, 132, 133
  lighting 137, 149
  nanocoatings 68, 72, 75, 78
  styrene production 94, 99, 105, 106
Lithography 4, 46, 52, 53, 124

Magic Nano 201
Management systems 12, 14, 182, 183, 203
Material efficiency 130, 136, 204
Molecular molding 46, 51, 52

Nanobionics 12, 186, 188, 190, 191, 194, 195, 198, 220
Nano-environmental engineering 187
Nanogovernance 200
Nano-Jury 202
Nanoparticles VII, IX, 2, 3, 4, 5, 9, 13, 15, 31, 39, 40, 48, 50, 51, 59, 67, 93, 95, 156, 158, 160, 161, 162, 163, 164, 165, 166, 167, 168, 169, 170, 172, 175, 176, 178, 179, 180, 183, 187, 194, 200, 201, 202, 205, 206, 207, 208, 209, 210, 213, 217, 218
Nanostructured surface IX, 15, 31, 95
Nanotubes 7, 8, 9, 31, 46, 94, 111, 161, 167, 170, 176, 178, 191, 207
 carbon 47, 66, 99, 105, 110, 118, 130, 136, 167
Naturalness 190
Nature 23, 24, 28, 31, 37, 42, 46, 53, 94, 118, 149, 162, 163, 175, 183, 185, 188, 191, 197
 learning from 189, 193
NGO 172, 199, 201, 205, 211, 213

OECD 75, 200, 211, 213

Path dependencies 183
Players 10, 12, 13, 14, 32, 33, 34, 36, 37, 181, 182, 183, 185, 189, 196, 197, 199, 200, 203, 220, 221
Precautionary options 10, 181, 182
Precautionary principle XV, 22, 23, 34, 175, 178, 206, 210, 216, 219
Precipitation 4, 46, 51, 156, 195

REACh 14, 34, 203, 208, 218, 219, 221
Regulation 11, 18, 21, 34, 35, 39, 149, 150, 167, 172, 179, 182, 203, 206, 208, 209, 210, 217, 218
Resource efficiency 7, 27, 37, 86, 186, 189, 215
Risk analysis 3, 30, 35, 202, 205, 218
Road maps 33, 184, 186, 195, 220

Self-organization 3, 4, 13, 15, 40, 41, 46, 53, 62, 63, 188, 189, 191, 194, 195, 218, 219
Self-organization principles 42, 191, 195
Self-replication 3, 13, 62, 63, 188, 194, 195
Self-reproduction 4, 15, 20, 60, 62, 63, 195, 219
Sol-gel process 46, 49, 50, 74, 78
Spider silk 190, 191, 192, 193
Standardization 184, 204, 211, 212
Suitability 189, 190
Sustainability VIII, 1, 5, 13, 14, 32, 37, 185, 187, 189, 221
Sustainability effects IX, 1, 17, 27, 65, 215
Sustainability guidelines 182
Sustainable nanodesign 204
Sustainable nanotechnology 10, 12, 181, 185, 186, 191, 220

TBT Agreement 210
Technological development VIII, 10, 11, 13, 14, 32, 33, 36, 38, 60, 95, 117, 179, 181, 184, 186, 190, 196, 198, 202, 216, 220
Technology characterization 13, 17, 23, 24, 25, 27, 30, 35, 38, 42, 206, 216, 217, 218
Technology push innovation 184
Titanium dioxide 6, 68, 161, 164, 168, 171, 172, 173, 174, 175
TSCA 207

UBA 66, 93, 204
Ultra-fine particles 31, 161, 164, 165

Woodrow Wilson Center 200, 212, 213

Printing: Krips bv, Meppel
Binding: Stürtz, Würzburg